# Access Networks: Technology and V5 Interfacing

For a complete listing of the *Artech House Telecommunications Library*,
turn to the back of this book.

# Access Networks: Technology and V5 Interfacing

Alex Gillespie

Artech House
Boston • London

**Library of Congress Cataloging-in-Publication Data**
Gillespie, Alex.
Access networks: technology and V5 interfacing/ Alex Gillespie.
p. cm.
Includes bibliographical references and index.
ISBN 0-89006-928-X (alk. paper)
1. Broadband communication systems. 2. Multiplexing. 3. Telecommunication—Switching systems 4. Telecommunication—Standards I. Title.
TK5103.4.G55 1997
621.382'16—dc21 97-7731
CIP

**British Library Cataloguing in Publication Data**
Gillespie, Alex
Access networks: technology and V5 interfacing
1. Computer networks 2. Interneworking (Computer networks)
I. Title.
004.6'7

ISBN 0-89006-928-X

**Cover design by Jennifer Makower**




**685 Canton Street**
**Norwood, MA 02062**



International Standard Book Number: 0-89006-928-X
Library of Congress Catalog Card Number: 97-7731

10 9 8 7 6 5 4 3 2 1

# Contents

# Preface

Access networks are the increasingly complex multiplexing and transmission systems that are being deployed between the users of the telecommunications network and the core switching systems. This book deals both with the technology of access networks and with their interface to the core network.

The first half of the book deals with transmission over the different physical media, ranging from simple copper pairs to solitons and quantum cryptography. It also deals with the more general issues related to powering, services, and asynchronous transfer mode (ATM). The second half of the book describes the V5 interface and goes on to introduce the broadband VB5 interface.

Access networks and V5 interfaces change the traditional roles of both operators and suppliers, creating opportunities for competition and allowing structural changes. The final result of these changes is still to be felt.

There are many people who have provided help and support during the writing of this book. Most important of all is my wife, Christine, who has shown outstanding tolerance during weekends and evenings for a number of years.

There are also a number of friends and colleagues who deserve thanks. The work on the V5 and VB5 interfaces has involved many people, most notably Geoff Harland, the ETSI SPS3 Working Party 3 chairman; Karl-Heinz Stolp of Deutsche Telkom, who initiated the V5 work; and my BT colleagues, Keith James and Mike Hale, who have also played leading roles and have reviewed many chapters. I would also like to thank my other colleagues at BT who have helped to review the work, in particular Peter Adams, Gavin Young, Dave Faulkner, Alan Quayle, and Adrian Manley. Many helpful comments were also received from the reviewers at Artech House.

I also wish to acknowledge my appreciation to BT for the opportunity to work on access networks and for the encouragement I have received during the writing itself. Any faults or errors are my own responsibility.

# 1 Introduction

"Begin at the beginning."

—Anonymous

New access technologies are changing the traditional means of access to telecommunications services. Not only do established operators have a wider range of technological options and potential suppliers, but new technologies and interfaces allow new operators to enter the markets as permitted by the regulatory authorities. This competition may also stimulate new services that can be supported by the new technologies. A dynamic evolution of access networks is promised with great benefits to the industry and its customers. This book describes the technologies and interfaces that are fueling these changes.

## 1.1 BACKGROUND

Access networks are the increasingly complex multiplexing and transmission systems that are being deployed between telecommunications users and the exchange switches of the core telecommunications network to which they are connected. They are a recent innovation, and most customers are still connected to their exchanges over passive copper pairs. The first steps towards access networks were taken with the introduction of remote multiplexers and pair gain systems. The trend towards more complex systems reached a critical point with the introduction of radio and optical-fiber technologies, which require even more sophisticated transmission systems.

The most important decision seems to have been made almost by accident. This was the decision to interface access networks to their host exchanges, rather than integrate them into these exchanges. Part of the reason for this was that many of the initial developments in optical access networks took place in the United Kingdom, where an open interface was available that permitted access networks to be developed independently of the exchanges.

As the concept of separate access networks became more widely accepted, operators in a number of countries came to realize that there was no suitable standardized interface available to connect these networks to their exchanges. In an attempt to solve this problem, DBP/Telekom took the initiative within the European Telecommunications Standards Institute (ETSI) to create a European standard, the V5 interface. This specification was subsequently submitted to the International Telecommunications Union (ITU-T) (formerly CCITT) and is now a global standard.

It would be nice to be able to attribute the development of access networks and the V5 interface to an enlightened desire to benefit mankind by enhancing the capabilities and reducing the costs of telecommunications systems. Unfortunately, this would fail to take proper account of human fallibility. Mankind will indeed benefit from these developments, but as much is owed to chance as to design.

## 1.2 STRUCTURE OF THE BOOK

This book is divided into two parts. The first part deals with the technological aspects of access networks, and the second part deals with the interface between access networks and the core telecommunications network.

The four chapters following this introduction cover media-dependent transmissions within an access network. Chapter 2 covers digital transmission over copper pairs, from basic rate *integrated services digital network* (ISDN) transmission at 144 Kbps to *very-high-speed digital subscriber loop* (VDSL) transmission at tens of megabits per second. Chapters 3 and 4 discuss optical-fiber technology. Chapter 3 covers the basic architectures, transmission techniques, and components, while Chapter 4 covers more advanced concepts such as optical amplifiers, coherent operation, new architectures, soliton transmission, and quantum cryptography. Radio technology and its application to access networks is covered in Chapter 5.

The two subsequent chapters deal with the wider technological aspects of powering and *asynchronous transfer mode* (ATM). The powering issues and possible solutions are covered in Chapter 6. Chapter 7 starts with a brief overview of ATM, and then focuses on ATM *passive optical networks* (PONs), addressing the issue of bidirection point-to-multipoint ATM transmission. The final chapter of the first half of the book, Chapter 8, covers the services that need to be supported by access networks and the social and psychological barriers that need to be overcome.

The second part of the book concentrates on the V5 interface, especially the multiplexing of signals and the nature of the V5 messages. This part is intended to give an introduction to the V5 interface, and it is meant to supplement, not replace, the detailed technical content of the standards. The second part should also provide an understanding of the principles behind the specification of the V5 interface. In some ways this is more important than the introduction to the interface, because there are lessons from the V5 interface specification that can be applied in other situations. In addition, there are peculiarities of the V5 interface that are caused by nontechnical factors and that should perhaps be avoided in the future.

The V5 interface, the physical structure of its links, and the associated services, ports, and architectural concepts are introduced in Chapter 9. The multiplexing used for the V5 interface is described in more detail in Chapters 10 and 11. Chapter 10 covers the physical and frame layer multiplexing and the format of the V5 message layer. It also introduces the convention used later in the discussion of V5 messages. Chapter 11 covers the frame relaying and multiplexing of ISDN signaling across the V5 interface.

The next two chapters describe the V5 control protocol and the V5 PSTN protocol which are relevant to both V5.1 and V5.2 interfaces. Chapter 12 gives the format, details and example message flows for both port control messages and for common control messages. Chapter 13 describes the requirements for a *public service telephony network* (PSTN) protocol and how they are mapped onto V5 messages, and includes an example of a very simple application.

The next three chapters describe the three protocols that are specific to V5.2 interfaces. Chapter 14 describes the *bearer channel connection* (BCC) protocol, which is used for the dynamic allocation and deallocation of channels on a V5.2 interface. Chapter 15 describes the link control protocol, which is used to take V5.2 links in and out of service and to check their integrity. Chapter 16 describes the protection protocol, which is used to improve the security of V5.2 communications channels by switching them onto backup time slots when problems arise. Chapter 17 describes the progress towards the broadband VB5 interface. Chapter 18 provides concluding remarks.

Most of the chapters of the book end with a summary so that the reader can quickly gain an initial understanding of the issues and concentrate further attention on those particular chapters of greatest initial interest. Many chapters also include a list of material for further reading so that topics of special interest can be investigated in greater detail.

# Advanced Copper Pair Technology

# 2

"The reports of my death are greatly exaggerated."
—Mark Twain

## 2.1 BACKGROUND

Most of the investment to date in access networks has been in the installed copper cabling used to provide PSTN service to homes and businesses across the world. Copper wires are twisted together to ensure that external electronic noise is picked up equally on both wires to form a longitudinal signal that can be removed by coupling to the wires with transformers. The pairs of wires are then spun into cables or into binder groups within cables to average out the effects of their physical position within the cables. Coaxial copper cables are used in some places because their external shielding improves their noise immunity, but they are larger and more expensive than copper pairs and so are mainly used in specialized applications, such as between multiplexers.

For a short time, aluminum wires were tried out instead of copper because an increase in the cost of copper was predicted. Aluminum wires need to be larger than their copper equivalent because of the higher electrical resistance of aluminum. In addition, aluminum is harder to work with because it is more difficult to solder. The predicted cost increase for copper did not materialize, so the use of aluminum was dropped due to the operational difficulties.

Paper insulation continued to be used on copper pairs even after the introduction of plastic insulation because the pressurization needed for paper insulation was still needed with the initial types of plastic insulation, and pressurized cables with paper insulation were still effective. The pressurization kept the paper dry by preventing water from entering the cable, keeping the copper insulation good. Pressurization was still needed with plastic insulation, since water traveling through pinholes in the plastic caused electrolytic corrosion of the copper. The pressurization could be removed when jelly-filled plastic cables were introduced because the jelly kept the water out without the need for pressurization. This removal of pressurization was disadvantageous in that the pressurization alarms which indicated the location of breaches in the cable were also removed. Nevertheless operating practices are simpler if cables do not need to be pressurized.

The performance of the copper pairs was sometimes improved by adding loading coils, small inductors, at various points along their length. This flattened the high-frequency end of the speech transmission spectrum by compensating for the capacitance between the pairs. Unfortunately this technique also prevented digital transmission on the pairs because the coils increased the reactance of the transmission path at the higher frequencies of operation needed for digital operation. These loading coils have now been removed in much of Europe, but they are still present in 25% to 30% of lines in the United States, and are especially common on longer lines. Lines in the United States often also have bridged taps, which are split ends introduced to create flexibility or as a result of disconnections and reuse of the pair. Bridged taps also impair digital operation, often disastrously, at a particular frequency.

Copper pairs generally consist of a number of lengths of wires of different gauges. Narrow gauge wiring is often used near the exchange when duct space is constrained. Wide gauge wiring is often used on long lines to keep the loop resistance down. These changes of gauge cause the reflection of digital signals, which if not corrected also seriously restrict bidirectional transmission on a copper pair.

Techniques have been used at times to extend the operation of the copper pairs. Where operation is limited by a dc feeding loop, extenders that boost the dc feed have been used. Where operation is limited by attenuation of the voice signal, then either hybrid amplifiers, which are matched to the line and boost both directions of transmission, or voice-switched amplifiers, which boost only the higher signal in a single direction, have been used.

The copper pair infrastructure is a major asset to the traditional operators of telecommunications networks. It allows considerable scope for the deployment of more advanced copper pair transmission technology to make more effective use of this asset.

## 2.2 BASIC DIGITAL TRANSMISSION

Although digital transmission over existing copper pairs has been widely used for fax communication, it falls far below the capabilities of the transmission medium because fax only uses the same frequencies as ordinary voice communications. At best, it can achieve about 28.8 Kbps, with 9.6 Kbps being achievable on most lines. Proper digital communication can achieve 160 Kbps over almost all copper pairs and can also be done at lower cost than with voice band communication because greater sophistication is required by modems to operate high rates over the narrow bandwidth of a voice-band link. Similar techniques of echo cancellation and equalization are used both for modems and for digital transmission, but the highest speed voice-band modems also use more sophisticated encoding and decoding to maximize the data rate over voice channels.

Digital transmission over copper pairs was developed to support basic rate ISDN operation of two B-channels at 64 Kbps each and one D-channel of 16 Kbps plus the associated synchronization and additional overheads. The technical challenge here is to achieve bidirectional transmission over almost all of the existing metal pairs. This is a challenge that has now been successfully met. Unfortunately, the ITU-T ducked the issue of defining a global standard for this transmission and produced a reference model and standards for other related interfaces (see Figure 2.1).

Part of the reason that the ITU-T did not produce a standard for the U-interface was the number of vested interests involved. Certain operators did not want a U-interface standard because its absence allowed then to make money on the rental of the electronics at the far end of the line. Suppliers also had vested interests, since many did not want a U-interface standard to be agreed on unless it took the form of the one they had already developed.

### 2.2.1 Techniques for the U-Interface

Three different approaches to duplex operation on existing lines have been considered. Two of these are based on a well-defined separation of transmit and receive directions, either in time or in frequency. The ping-pong or *time-compression multiplexing* (TCM) approach allows a copper pair to be used by each end at certain times for transmission and at other times for reception (see Figure 2.2). This requires a large bandwidth because each direction can only use the pair for less than half of the time, since it is necessary to wait after transmission for echoes of the transmitted signal to die out before signals from the far end can be received.

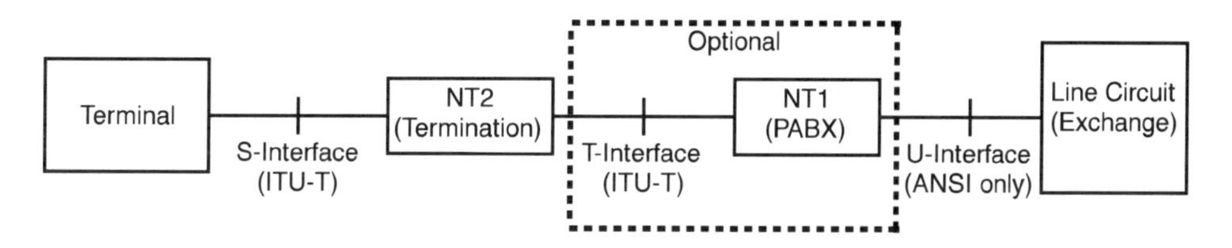


**Figure 2.1** Interface standards for basic rate ISDN transmission.

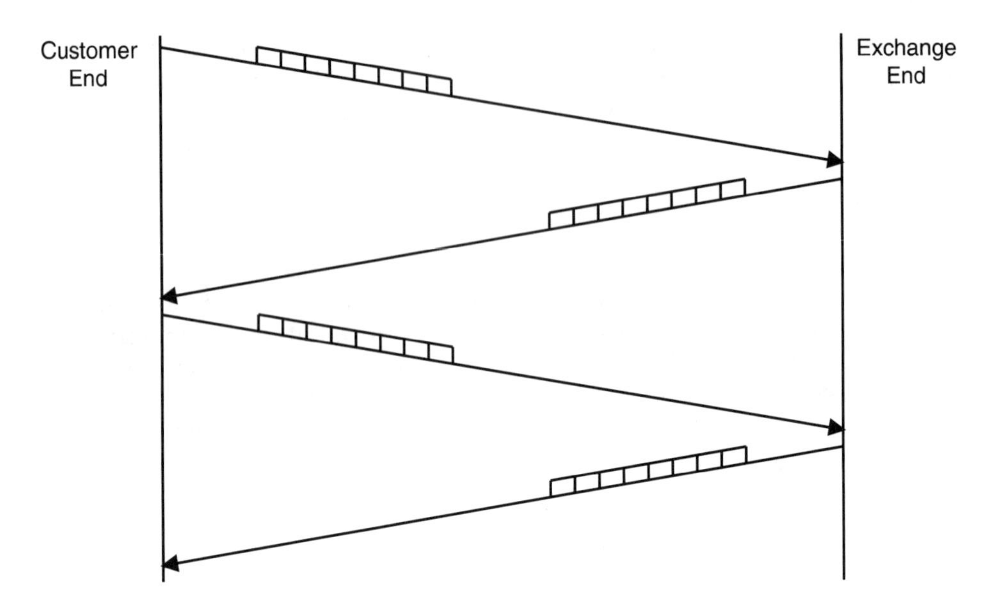


**Figure 2.2** Ping-pong or time-compression multiplexing (with one "ball").

There is a common fallacy about ping-pong operation. It has often been stated that its range, which is limited by attenuation, is also limited by the round-trip transmission delay. This idea can be visualized by representing the bursts of data on the line as a ping-ping ball that takes time (about 5 μs per kilometer) to travel from one end to the other. It is commonly said that the range is limited because the "ball" must be returned before another burst of data can be sent; that is, the frequency of the bursts is limited by the round-trip delay of transmission. To some extent, this can be overcome by increasing the size of the ball (i.e., by putting more information into each burst), but this approach is also limited because it increases the end-to-end delay, since a burst must be filled before it is sent. If the size of the ball is limited and the frequency with which it is sent is limited, then the fallacy is to conclude that the throughput is limited.

This is a fallacy because it assumes that the game is only played with one ball. Playing with two or more balls is more difficult, but the added complexity of a transmission system based on this approach is far less than that for a system based on the American National Standards Institute (ANSI) standard. The true limitation to ping-pong operation is due to the increased attenuation and interference arising from the higher bandwidth needed as less time is available for transmission in either direction. This makes the system suitable only for operation on shorter lines, where the simplicity of its implementation is a major advantage. It has been very popular in Japan, where the limited line length is less important because of the local geography.

The separation of directions of transmission by frequency requires a similar bandwidth to that of the ping-pong approach, which uses separation in time. In both cases, the fundamental bandwidth needs to be doubled. The additional margin required for implementation of the filters of the frequency separation technique is balanced by the additional bandwidth required for ping-pong operation to allow the echoes to die out. The ping-pong technique, however, is simpler to implement because it is purely digital and does not require the sharp analog band-pass filters.

The ANSI standard calls for a third technique, echo cancellation, because it has a lower bandwidth than either time or frequency separation and so can be used on longer lines.

### 2.2.2 Echo Cancellation and Equalization

Time and frequency techniques distinguish explicitly between the two directions of transmission. The technique of echo cancellation allows the received signal to be calculated by subtraction if both the line characteristics and the transmitted signal are known.

If the output impedance of the transmitter matched the complex impedance of the terminated line, then the signal on the line would be exactly one half of the transmitted signal (see Figure 2.3). The signal received from the other end of the line could then be obtained by subtracting one half of the output of the transmitter from the signal on the line. Unfortunately, the impedance of the

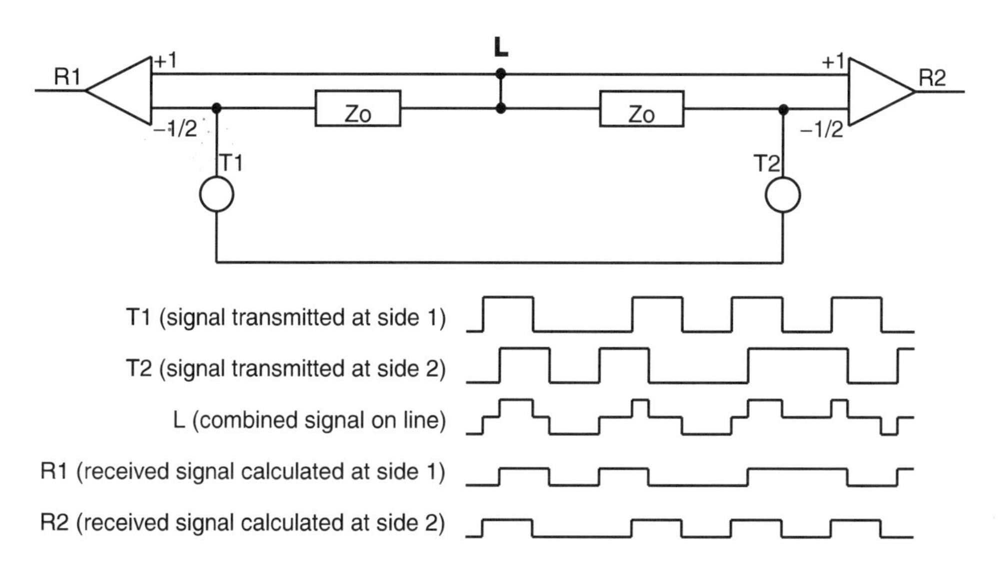


**Figure 2.3** Calculation of received signal by subtraction of transmitted signal.

terminated line is complex and varies from line to line, so a received signal that is extracted simply contains echoes of the transmitted signal.

These echoes are due to the mismatch between the matching impedance and the characteristic impedance of the line and between the different characteristic impedances of the different sections of the line. The echo due to the mismatch between the characteristic impedance of the final section and the terminating impedance at the other end is not as significant because that echo is smaller than the signal that is sent from the other end. Echo cancellation works by subtracting out a signal obtained by adaptively estimating the echoes produced by these mismatches (see Figure 2.4).

For echo cancellation to be successful, it is necessary that there be no correlation between the transmitted and the received signals. If this condition is not met, then the received signal may resemble an echo of the transmitted signal and the echo canceler may attempt to cancel the received signal because it thinks it is an echo. To ensure that there is no correlation, a different scrambling algorithm is typically used at each end of the line to reduce the probability of correlation occurring by random probability.

Implementations of echo cancelers need to take account of nonlinearities of components and of the long tails that echoes may have. Basic echo cancellation often makes use of filters with a *finite impulse response* (FIR), which are linear and cease to cancel after a finite time. Long tails may need to be canceled using *infinite impulse response* (IIR) filters, which are linear and typically have an exponentially decaying tail. Nonlinear effects, due to amplifiers or transformer saturation, typically require a memory-based cancellation (i.e., a look-up table that adaptively estimates the nonlinear corrections for the particular line and line circuit). Memory-based cancellation is theoretically the best approach because it makes no assumptions about the nature of the echoes, but the memory requirements for normal implementa-

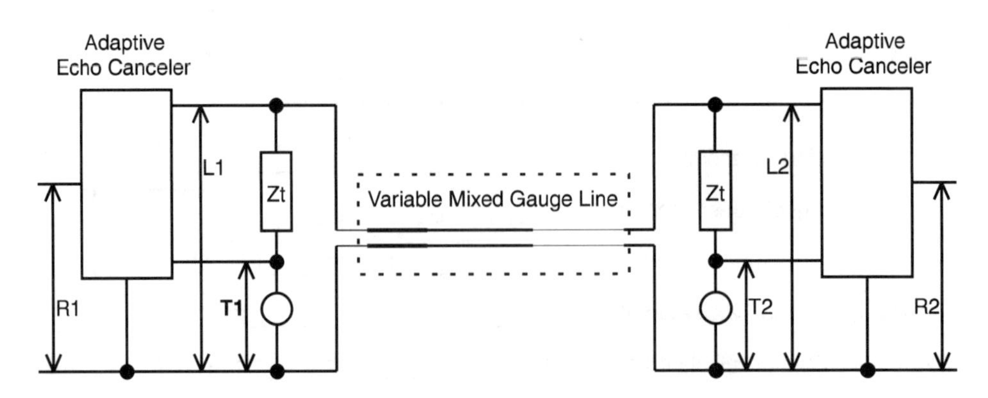


**Figure 2.4** Adaptive echo cancellation.

tions grow exponentially with the length of the tail, making it impractical in most cases.

The design of echo cancelers is also complicated by the digital phase-locked loops which are normally used to recover the clock from the received data. Digital phase-locked loops are often used because they are easier to implement in a digital chip design. Unfortunately, they make the echo harder to estimate because the point where the received signal and echo are sampled jumps as the loop adjusts to a different number of step sizes. This may require an interpolation to be made between two estimates of the echo to avoid sudden changes due to this design limitation.

The received signal is also distorted by transmission down the line. A pulse that is transmitted from one end is often dispersed over several pulse periods by the time it reaches the other end of the line. Typically, the received pulse contains a long trailing tail, which interferes with subsequent received pulses, creating what is known as *intersymbol interference* (ISI). Like the echoes due to impedance mismatches, the intersymbol interference can be eliminated if the characteristics of the line are known. A filter can be added to equalize the delays and attenuations of the different frequency components of the pulse (see Figure 2.5). Typically, this filter also needs to adapt to the specific characteristics of each particular line, although under certain special conditions a fixed equalization filter can be very successful. After a received pulse is decoded, any remaining tail can be canceled using the techniques that correspond to those used to cancel the effect of echoes. This corresponding approach is known as *decision feedback equalization* (DFE) because after a decision is made on the pulse, the result is fed back to assist the equalization by removing the tail that ought then to follow. DFE has a disadvantage in that each time a pulse is decoded incorrectly, it increases the probability

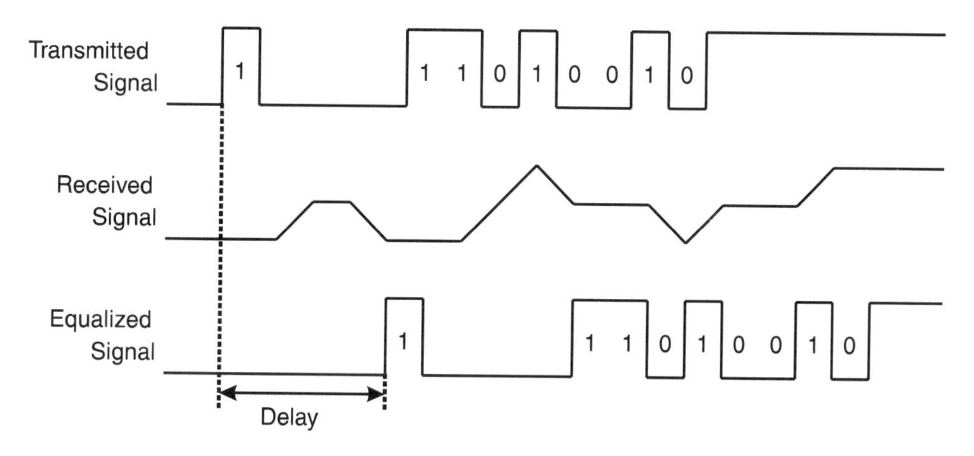


**Figure 2.5** Equalization of intersymbol interference.

that the next pulse will also be decoded incorrectly because a false estimation of the tail will be subtracted.

### 2.2.3 Line Codes

The main distinction between the different schemes considered for the ANSI standard were the codes used to represent digital information for transmission—the line codes. The final decision, 2B1Q, has the merits of both technical superiority and of not giving an advantage to any particular supplier. Its technical superiority is a result of having the lowest bandwidth requirement and so suffers least from attenuation and noise.

The 2B1Q code represents pairs of bits (2B) as a single four-level value (1Q). Alternatives to this typically use three-level (ternary) codes. The 3B2T code represents sets of 3 bits (3B) with eight possible combinations as pairs of ternary values (2T), allowing nine combinations that can be reduced to eight if, for instance, the ternary pair 0-0 is not used. Similarly, the 4B3T code represents sets of 4 bits (4B) with 16 possible combinations as sets of three ternary values (3T), permitting 27 combinations. The 4B3T mapping can be reduced to two 3B2T mappings if the first of the 4 bits to be mapped determines whether the first of the ternary values is +1 or −1 and the remaining 3 bits are mapped according to 3B2T, suggesting that 3B2T may have better performance than 4B3T, since its baud periods are slightly larger. Despite this, 4B3T gained a greater popularity, due in part to commercial support. The spare combinations in 3B2T and 4B3T codes may be used for special functions or alternatively to improve the spectrum of the codes or the performance in the presence of noise. The various codes are illustrated in Figure 2.6.

The simplest ternary code is *alternative mark inversion* (AMI), which represents alternative binary 1s as +1 and −1. This has the disadvantage that clock recovery can be impaired if a long string of 0s is transmitted. To assist clock recovery, the data is typically scrambled, but this does not introduce additional complexity because scrambling is required in any case for the echo cancellation.

One of the simplest codes is the two-level biphase code. In simple terms, this can represent a 1 by a positive phase transition at the center of a bit, and a 0 by a negative phase transition. To avoid the need to label the separate strands of a copper pair, which creates operational problems in the field, it is more sensible to use differential biphase coding. This represents a 1 as a single square wave and a 0 as a half cycle of a square wave with twice the period, and there is also a zero crossing at every bit boundary.

The disadvantage of the biphase coding is that it requires about twice the bandwidth of most other codes, but it has a compensating advantage that a very simple implementation is possible. Because the bandwidth is high and the spectrum has little energy at low frequencies, the echoes die out quickly,

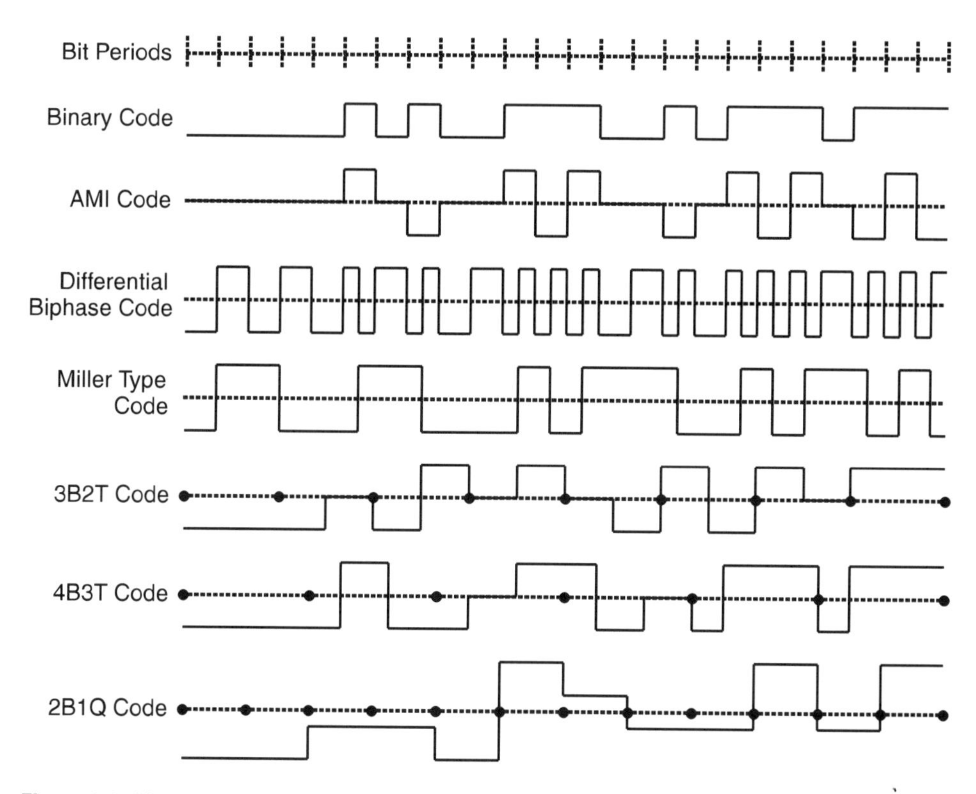


**Figure 2.6** Line codes for transmission over copper pairs.

making a memory-based echo canceler feasible. In addition, an implementation can make do with a fixed equalizer because the code is partially self-equalizing. Self-equalization occurs because the dispersion of the 0s and 1s can balance out over a long length of lines, since coding the 0s as a half cycle with a large third harmonic content produces a signal with similar properties to 1s encoded as a full cycle. Biphase coding is closely related to Miller codes, which have a much smaller spectrum. For example, one type of Miller code represents a 1 as a midbit transition and a 0 as no midbit transition and introduces an end-bit transition after two successive 0s if they are followed by a third 0. Using a Miller code instead of a biphase code creates the possibility of lower spectrum code, which also leads to simple implementations, since the absence of energy at low frequencies again causes the echoes to die quickly.

In contrast, the 2B1Q code chosen by ANSI has one of the most complex implementations. It requires both adaptive equalization and echo cancellation, and the echo cancellation may need to use several techniques in conjunction because of nonlinearities and the long time required for echoes to decay. This technical complexity has resulted in additional implementation delays.

The complexity of the implementation of the ANSI standard raises the issue of why only a single code was chosen. The cost and development delays could have been reduced if the agreed-on ANSI standard were used for long lines and a simple approach, such as biphase or ping-pong which had already been developed, were used for shorter lines. It is curious that in this area, as in so many areas of standards activities, a complex solution is chosen in preference to a simple solution. It may be unreasonable to expect a simple solution to be found, since simplicity is often obvious only with hindsight. However, it would be very uncharitable and certainly incorrect to suggest that standards experts are uneasy with simple solutions because simple solutions provide less scope for the demonstration of their professional expertise.

### 2.2.4 Noise and Impairments

Noise is normally the limiting factor in digital line transmission because the signal processing techniques described above can compensate for the echoes and the intersymbol interference produced by real lines. The ability to operate at an acceptable error rate is determined by noise, which becomes more significant as the signal is attenuated by transmission down a line.

The dominant noise for basic digital transmission has two components: *near-end crosstalk* (NEXT) and impulse noise. Crosstalk (see Figure 2.7) is produced by unbalanced couplings between different pairs in a cable. When the couplings are unbalanced, the signals from adjacent pairs cause a differential signal across the two legs of a pair because both legs are not affected equally. The component of the crosstalk signal that continues to travel down the cable in the same direction after coupling into an adjacent pair is known as *far-end crosstalk* (FEXT). The component that propagates back up the cable towards the transmitting source after coupling into the adjacent pair is NEXT.

NEXT has a greater impact than FEXT for symmetric bidirectional transmission because the FEXT is attenuated both by the crosstalk coupling and by the transmission down the full length of cable, whereas NEXT need only travel

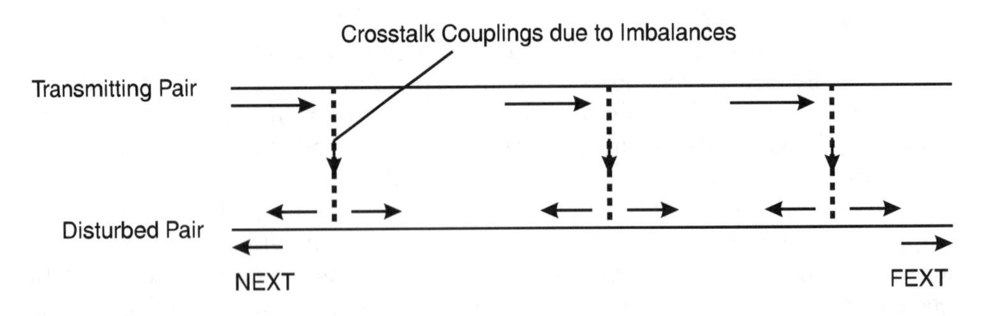


**Figure 2.7** NEXT and FEXT.

a small distance and then back again. The NEXT from different adjacent pairs is normally treated as if their phases are random; that is, the total power due to crosstalk averages out to the sum of the power from every coupling. This is an oversimplification because the crosstalk noise from imbalances that are nearby tend to be larger, since they have been attenuated less by transmission, and the overall result tends to be either in-phase or 180 deg out-of-phase depending on which leg of the pair receives the most signal. This means that the total NEXT that couples into a pair is slightly larger than would be estimated by adding the individual powers. Even so, for most environments, the crosstalk noise is not as great as the impulse noise.

Impulse noise is due to electromagnetic interference that comes from a variety of different sources. One of these sources is noise in exchanges. Older electromechanical exchanges may be the worst offenders because of the pulses produced by their coils, but modern digital exchanges also produce noise which has a greater effect than might be expected, since it is synchronized to clock signals. Impulse noise is also generated when ringing voltage is switched on and off, when the line polarity is reversed, and when the handset on an adjacent pair goes off-hook or produces make/break signaling pulses. Other electrical equipment may also produce impulse noise.

Digital transmission can also be significantly impaired by bridged taps, which are common in line plants in the United States. Bridged taps are additional pairs which are connected to an existing pair to provide flexibility (see Figure 2.8). Typically, a bridged tap is not terminated at its remote end. Under certain conditions, a bridged tap can kill signal transmission. In particular, if an unterminated bridged tap has a length corresponding to a quarter of the wavelength of a signal being transmitted, then the reflection from its far end cancels the signal being transmitted down the pair. No amount of simple signal processing can compensate for the characteristics of a line with such a "killer" bridged tap because the line behaves as if there is a localized short-circuit at

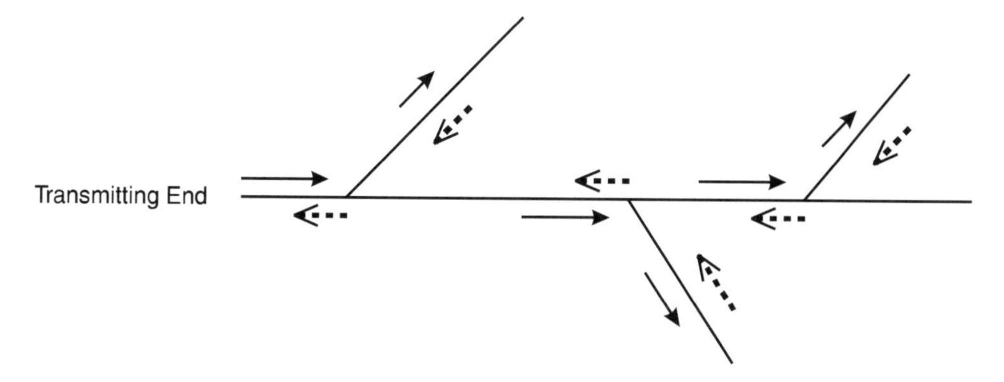


**Figure 2.8** Reflections from junctions and ends of bridged taps.

that frequency, making transmission at that frequency impossible. However, a decision feedback equalizer can allow transmission even in this case because the absence of a signal is information in its own right, and the decision feedback equalizer can add back the signal that has been canceled by the reflection.

Under certain circumstances, a signal can be transmitted on a line with a killer bridged tap if a second bridged tap is added because of the multiple signal reflections introduced by the additional bridged tap. There are other strange wave phenomena that can also occur due to interference over a relatively short length of copper pair. It is even possible to have lower attenuation when a short length of an unmatched pair is added because it can act in the same way as antireflecting coatings on camera lenses and reduce the signal attenuated by reflection.

Digital operation on certain lines is simply impossible because they have had loading coils added to improve their performance over the voice band. Loading coils are simple small inductors added at intervals to modify the characteristics of the line at voice frequencies by offsetting its capacitance. These loading coils prevent digital transmission because they give the line a very high impedance at the frequencies of operation.

### 2.2.5 Concluding Remarks

Digital transmission capable of supporting ISDN basic rate traffic is possible over almost all of the existing copper pairs in the access network. The U-interface chip sets cannot achieve their theoretical line lengths in every case because of noise and impairments. In particular, the worst-case impulsive noise in real systems and the presence of killer bridged taps and loading coils can prevent digital operation.

With hindsight, it now appears that it might have been possible to achieve adequate performance with less sophisticated techniques on the majority of subscriber lines, while even more sophisticated approaches than those initially developed are needed to avoid the use of repeaters on the most difficult lines.

## 2.3 PAIR-GAIN SYSTEMS

The traditional use of pair-gain systems has been as a temporary and expedient solution to the problem of providing service when there are no spare pairs in a copper cable or when there is no spare room in a duct for a new cable. The initial types of pair-gain systems supported a second, nontraditional PSTN connection on higher frequency carriers in addition to the normal traditional connection. In some cases, these pair-gain systems have been used as a permanent solution despite their high cost because the cost of installing a new duct or a new cable is even higher and is not justified by revenues.

More generally, a pair-gain system is a form of multiplexer or concentrator that is used between a customer and the exchange to which the customer is connected. In the present context, only pair-gain systems operating over copper pairs will be considered, because they are likely to become more significant as operators become more reluctant to install new copper pairs, either because customers are being lost to competitors or in anticipation of installing a fiber infrastructure. With fiber and radio systems, it is also a bit pointless to talk about pair-gain systems because there are no physical pairs and because multiplexing is the norm. However, certain of the techniques used in copper-pair-gain systems can also be applied to fiber and radio systems.

A pair-gain system can use analog or digital transmission, or it can be a hybrid system that uses both. Older systems used analog transmission, while more recent systems use digital transmission. If a single, nontraditional PSTN connection is used in addition to the normal PSTN connection, then the system is designated as a 1+1 system. There are also 0+2 and 0+4 systems, which have no traditional PSTN connections and have two or four nontraditional connections, respectively. Both analog and hybrid 1+1 systems have been developed, but a digital 1+1 system is a contradiction, since the traditional PSTN connection is analog. The hybrid 1+1 system needs special filters to prevent the switching of high voltages used for analog signaling on the traditional PSTN connection from interfering with the digital transmission. These pair-gain systems are illustrated in Figure 2.9.

Pair-gain systems using analog transmission have also been developed for 0+8 operation, but these are increasingly more expensive than systems using digital transmission because of the higher cost of traditional analog filters. These could be redesigned using *digital signal processing* (DSP) technology, but

| Technology | Traditional PSTN | No traditional PSTN | | |
|---|---|---|---|---|
| Analog | 1+1 | | | |
| Hybrid | 1+1 | | | |
| Digital | | 0+2 | 0+4 | 0+... |

**Figure 2.9** Pair-gain systems.

this is unlikely because of recent advances in high-speed digital transmission over copper pairs.

Pair-gain systems using digital transmission have the added advantage of being able to also use compression techniques to reduce the bandwidth required for PSTN. Although uncompressed PSTN requires a bandwidth of 64 Kbps, it is relatively simple to compress speech into both 32 and 16 Kbps. The digital 0+4 system that has been developed uses ISDN U-interface technology to support 128 Kbps plus associated signaling, with 32-Kbps *adaptive delta pulse code modulation* (ADPCM) technology to give four PSTN connections. The 0+2 systems are similar, but do not use ADPCM. The 16-Kbps ADPCM technology is less appropriate because it introduces coding delays that reduce the quality of service and because it causes greater distortion when high-speed modems are used on the connection.

Digital pair-gain systems can make use of *high-speed digital subscriber loop* (HDSL) technology or more traditional techniques of supporting digital transmission at 1,544 or 2,048 Kbps. At these rates, two copper pairs can support up to 30 PSTN connections without compression, or 60 PSTN connections if 32-Kbps ADPCM is used. The technology used for HDSL is described in more detail later in this chapter; however, this level of pair gain creates significant problems for powering.

Analog concentrators have also been used as pair-gain systems, particularly in low-calling-rate residential areas. These systems take advantage of the low probability that more than a minority of the far ends will be active at any time. Concentrators use a small number of pairs to support a high number of potential connections with pairs being allocated to users on demand. Concentrators are characterized by the total number of lines supported and by the number of lines used to do this. A 14/5 concentrator supports a total of 14 lines using five actual lines to do so.

In addition to 14/5 concentrators, 15/6, 90/16, and 160/28 concentrators have also been developed (see Figure 2.10). As the total number of supported lines increases, proportionally fewer actual lines are needed, because statistical fluctuations in the use of lines are averaged out. As the size increases, the level of concentration approaches the ratio 8:1, which is used within exchanges to concentrate traffic from customers who are directly connected. Since both exchanges and pair-gain systems concentrate traffic, it is important to ensure that there is no combination of concentration in a pair-gain system and in an exchange, which would reduce the quality of service to an unacceptable level.

Concentrators may use either multiplexed or unmultiplexed transmission. If two copper pairs are used with digital 0+4 transmission on each, then a plausible concentration ratio of about 4:1 (see Figure 2.10) would allow a total of 32 PSTN lines to be supported. If HDSL transmission was used on these two pairs, then potentially the 60 links available with 32-Kbps ADPCM could be

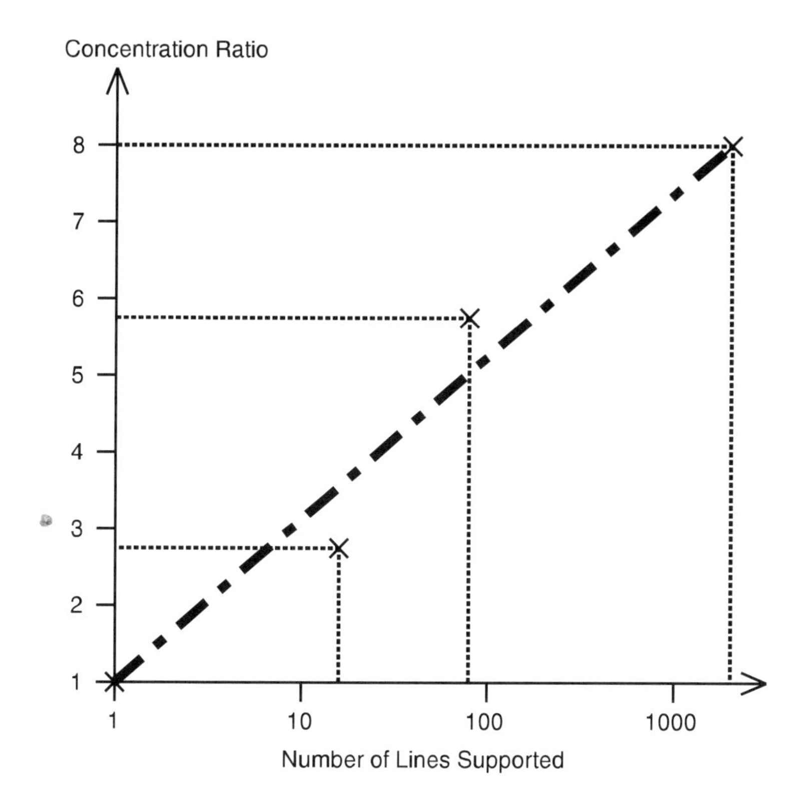


**Figure 2.10** Concentration ratio against size.

used to support 480 lines because the size of the system makes a concentration ration of 8:1 possible. Although this example is extreme, the possibility of supporting 480 PSTN customers on two copper pairs indicates the potential of pair-gain technology.

Concentrators that do not use multiplexed transmission can be relatively easy to power because there are sufficient lines used to power the supported lines that are active. This advantage can be lost if the normal human tendency to add complexity such as sophisticated signaling systems increases their power consumption.

Pair-gain systems that do use multiplexed transmission often require special powering arrangements. Even the simplest analog 1+1 systems use a trickle-charged battery to buffer the demands for power at the far end, and this has caused operational difficulties where there is heavy usage. The more recent digital 0+2 and 0+4 systems have related problems. The problems of powering remote nodes are discussed in detail in a separate chapter.

In the future, pair-gain systems may become more sophisticated as new capabilities are added. There is already a limited switching ability in some systems, and it has been suggested that these should be enhanced to create microexchanges. It has also been suggested that new features may be added more easily if they are included in pair-gain systems on the periphery of a telecommunications network, rather than at the core. In addition, there are plans to include greater fault location and testing capabilities into pair-gain systems. The last option may be the most sensible because it may help to offset the increased operational costs that the deployment of pair-gain systems will produce. The creation of complex pair-gain systems with these increased capabilities may be in the interests of telecommunications suppliers, but it is less clear that this is in the best interests of operators who wish to reduce their operational costs.

The role of pair-gain systems in the future is likely to increase in significance because of the deployment of optical fiber and because of increased competition between operators. It is not sensible for existing operators with extensive copper pair networks in place to upgrade the copper pair infrastructure if it is believed that optical fiber will be the preferred medium in the long-term future. Where additional capability is required, pair-gain systems are likely to be more extensively used. Pair-gain systems also allow new operators to compete with existing operators without the need for, or delay associated with, the installation of an extensive copper pair infrastructure. In particular, new operators may be able to support a large number of customers by using pair-gain systems if they can rent copper pairs or HDSL transmission from traditional operators.

## 2.4 HIGH-SPEED DIGITAL SUBSCRIBER LOOP

Digital transmission at 1.544 and 2.048 Mbps was developed before transmission at 144 Kbps for basic rate ISDN. The digital subscriber line transmission techniques subsequently developed for ISDN U-interfaces have led to innovations in the transmission at higher rates because these higher rate transmission systems have used older techniques. The application of the new approaches developed for basic rate ISDN transmission to higher transmission rates has led to the development of HDSL technology.

### 2.4.1 Requirements for HDSL Technology

The commercial incentive for the development of HDSL technology came from the United States, where cost savings were foreseen if the repeaters of the conventional T1 systems operating at 1.544 Mbps could be eliminated. An advantage was also foreseen if the new systems would operate on most copper pairs, and not just those which were specially selected for T1 transmission. At the time it was also believed that HDSL systems would have to be developed quickly because they were seen as a transitional technology with a window of opportunity that would close with the widespread introduction of optical fiber. A cynic might be tempted to think that this last argument had merit in that it would induce the development to be funded quickly and would allow copper pair technology to share some of the glamour associated with optical fiber, but surely no professional expert who had spent years developing copper pair technology would be this cynical.

The customers who require T1 systems for their services are predominantly business customers. For this targeted market, the requirement for line lengths can be relaxed to less than that necessary for basic rate ISDN line transmission because most business customers are located relatively close to the their telephone exchanges. The *carrier serving area* (CSA) specified in the United States for HDSL technology is 12 kft of 24-gauge cable or 9 kft of the higher resistance 26-gauge cable. These lengths appear reasonable because, for example, a survey failed to find any length in the Manhattan area that was longer than 10 kft and because long lines with loading coils are excluded. The specified CSA may prove impossible to meet in all cases because many lines have bridged taps, including the killer bridged taps described earlier for basic digital transmission.

The initial emphasis on 1.544-Mbps HDSL technology is reflected in the European developments for 2.048-Mbps operation. The traditional systems for 1.544-Mbps transmission used two pairs—a transmit pair and a receive pair. The corresponding HDSL systems also use two pairs, but use a dual duplex approach at 784 Kbps on each. When this technology is adapted to the 2.048-Mbps rate, three pairs are needed (i.e., a triple duplex approach). The alternative is to use dual duplex at a higher rate with a corresponding reduction in line length. This is illustrated in Figure 2.11.

| HDSL rate | No. of pairs | Pair rate |
|---|---|---|
| 1.544 Mbps | 2 | 784 Kbps |
| 2.048 Mbps | 3 | 784 Kbps |
| | 2 | 1176 Kbps |

**Figure 2.11** Options for HDSL transmission.

### 2.4.2 Techniques for HDSL Operation

The fundamental innovation that makes HDSL technology feasible is the introduction of echo cancellation. The techniques for echo cancellation were refined for use in basic rate ISDN technology, where duplex operation on a single pair is required. By using duplex operation on both the pairs that would traditionally be required for 1.544-Mbps operation, the operating rate of each pair is approximately halved. This is a significant advantage because the limit of operation in the absence of external noise is determined by the attenuation due to transmission and by the level of the NEXT, both of which are reduced when operation is at lower frequencies.

Applying echo cancellation at a higher rate requires the DSP to work faster because the signals change more quickly. In addition, more DSP taps are required because the tails of the signals are relatively longer due to the line characteristics. These differences do not present fundamentally new challenges. The echoes that need to be canceled still have the same sources: changes of gauge, mismatch of the terminations, and bridged taps.

There are some modifications to adaptive equalization when the techniques used for *digital subscriber loop* (DSL) transmission are extended to HDSL transmission. The line characteristics are different because at very high frequencies the line resistance is dependent on skin effect, which is a result of current being confined to the skin of the copper wire due to the conductor attempting to exclude changing magnetic fields. Depending on the gauge of the wire and on the line coding, the frequencies of operation may include the transition region from normal to skin-effect dominated transmission. These changing line characteristics do not require a fundamental change in the equalization, but improving the noise tolerance can.

In basic digital transmission, there may be fixed equalization in the transmitter, but equalization is normally carried out in the receiver because the receiver is able to assess how the received signal has been distorted. Normally DFE is used, with the knowledge of the detected pulse being fed back to cancel the tail of that pulse that extends into subsequent bits. DFE results in the propagation of errors at the limits of operation, even if the equalizer's coefficients are perfect, because a pulse that is wrongly decoded produces an erroneous cancellation in subsequent periods. Thus, a single error increases the probability of a subsequent error and so can produce a chain reaction.

This propagation of errors can be avoided if DFE is not used in the receiver. A more sophisticated equalizer would delay decoding the pulse until all of the information contained in the tail of the pulse had been received, but this can still result in error propagation, since assumptions need to be made about the nature of the subsequent pulses, which could need to be revised after those pulses had been decoded—and so on indefinitely. In addition, more information needs to be processed in the equalizer, making the implementation more complex than necessary. The source of the problem is not any errors in the equalizer's coefficients, but the uncertainty in the transmitted data. The most effective way to avoid the propagation of errors in the receiver is to move the equalization into the transmitter, because in the transmitter there is no uncertainty about the transmitted data.

The first difficulty with adaptive pre-equalization of the transmitted signal is that the transmitter does not know how signals are distorted by the time they reach the receiver. This can be solved either by making an intelligent assumption or by feedback. The intelligent assumption approach is for the transmitter to assume that the channel is symmetric in both directions and to use its knowledge of the distortion on the returned channel to pre-equalize the transmitted signal. This assumption is a good first approximation, but is questionable because the channel is often asymmetrical due to changes of gauge and to the higher probability of bridged taps at the customer's end. It also needs to be refined if adaptive pre-equalization is used at both ends, because then the remote end will act to eliminate any distortion on the returned channel.

The alternative to the assumption of a symmetrical channel is for the receiver to feed back information on the channel to the transmitter. This can be done initially on the basis of probing signals transmitted on power-up, which are analyzed by the receiver and then fed back to the transmitter using a low transmission rate, which can be easily achieved. Following this initial training-up, the receiver can continue to analyze incoming signals and inform the transmitter of the results during operation at the normal rate.

The second challenge with adaptive pre-equalization of the transmitted signal is that the adaptive system must be stable and produce a signal of limited amplitude. Instability has to be avoided because there is a propagation delay from transmission down the line, and an attempt to equalize out this delay using feedback from the far end would produce positive feedback and oscillation. This is not a problem if the equalizer is in the receiver, because the operation of the equalizer is independent of true propagation delay, since it has no knowledge of it. If the equalizer is in the transmitter, then the feedback from the receiver can result in an unstable loop. For the equalizer in the transmitter to be stable, it must combine with the line characteristics to produce a uniform delay across the spectrum of the pulse (see Figure 2.12).

It is also necessary to ensure that the pre-equalized output of the transmitter does not exceed the maximum signal level its electronics can handle, and this can be achieved using Tomlinson precoding. Tomlinson precoding works by using modulo arithmetic in the DSP of both the transmitter and of the receiver. A basic pre-equalizer takes the output signal and combines it with the input signal using a model of the line, so that when the actual line is added the result is a delayed version of the input signal. Applying modulo arithmetic at the output of the transmitter is equivalent to adding a fixed offset to the input of the precoder, which prevents the output from going out of range. The

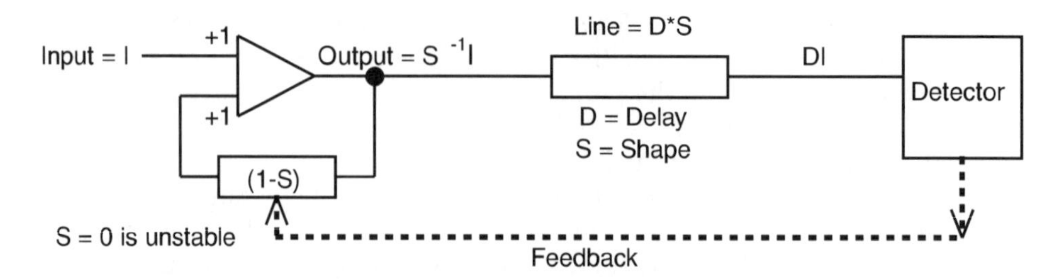


**Figure 2.12** Feedback for adaptive pre-equalization.

Tomlinson precoding and the line transmission balance out to reproduce the input plus the fixed offset at the receiver, which then removes the fixed offset by taking the modulus of the received signal (see Figure 2.13).

The use of modulo arithmetic in Tomlinson precoding is similar to the use of modulo arithmetic in trellis coding schemes, where the information in a transmitted baud is deliberately spread out over several bit periods to create artificial intersymbol interference, and modulo arithmetic is used to combine the contributions due to the individual bits (see Figure 2.14). In this simple case, if the binary code is 1, then the next bit is coded as a positive transition if it is also a 1, or as a negative transition if it is a 0. If the binary code is zero, then the next bit is coded normally. This coding is equivalent to introducing intersymbol interference, which extends the bit period by a half and adds the tail of the previous bit to the current bit using modulo 2 arithmetic. Although the bandwidth is increased, the transmission can tolerate more noise because a 1 is confirmed by a transition in the next bit and a 0 is confirmed by the absence of a transition. Artificially introduced intersymbol interference of this kind, which is corrected by suitable signal processing in the receiver, could also be used to improve the performance of HDSL systems. The use of this or other forms of error correction coding is improved if the equalization is performed in the transmitter, because these error correction techniques, like equalization, are prone to error propagation. If the equalization does not cause errors to propagate, then there is greater scope remaining for error correction coding because errors are not propagated twice.

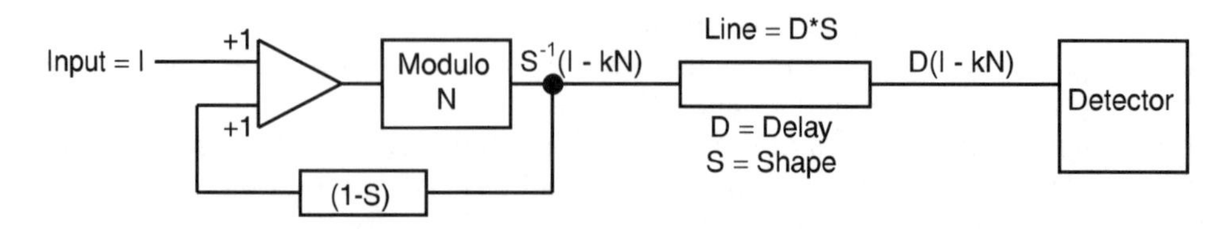


**Figure 2.13** Operation of a Tomlinson precoder.

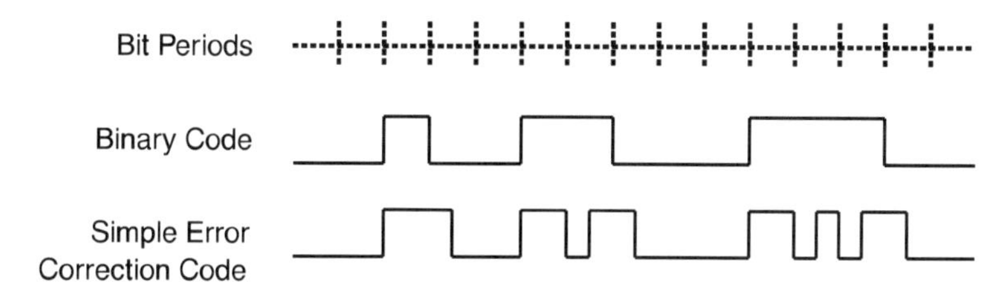


**Figure 2.14** Error correction coding through artificial intersymbol interference and modulo addition.

The performance of HDSL systems can also be improved by taking advantage of knowledge of the known properties of NEXT. The simplest way this can be done is by compensating for the fact that NEXT does not have a white spectrum. This can result in an improvement of about 1 dB in the signal-to-NEXT ratio.

A much greater improvement, estimated at about 6 dB, is achievable because it is theoretically possible to adaptively cancel all of the NEXT due to transmission to the customers. The information being transmitted at the exchange on all pairs in a cable can be made available to all receivers at the exchange end, and the receivers can use it to cancel the NEXT using echo cancellation techniques, since NEXT is an echo of the signals transmitted on other pairs. This concept is unlikely to be of practical use, because the complexity of the implementation is likely to result in it not being developed until after advances in optical fiber deployment make it redundant. It also suffers from the disadvantage that it does not overcome the NEXT limit on reception at the customer's end, and the problems of impulsive and other noise still remain unsolved by this approach.

A potentially more viable approach is to compensate for the NEXT between the pairs of an HDSL system. This type of NEXT can also be viewed as a special type of echo that can be canceled as there is knowledge of the signal that has produced the echo. If this NEXT is not canceled, then it is likely to limit performance because the pairs are likely to be close together in the cable and so have strong crosstalk couplings between them. This approach, known as *coordinated transmission*, appears capable of giving several decibels of improvement. It does not suffer from the implementational complexity of the previously described technique and it reduces NEXT at both ends of the line, since the signal transmitted on both pairs is known. It also has the potential advantage that its use can be further extended to allow the signal-to-noise ratio to be averaged over the pairs if independent subchannels are transmitted simultaneously with suitable weightings over each pair, rather than being limited by the worst case.

### 2.4.3 Performance of HDSL Systems

NEXT imposes a fundamental limit on the performance of HDSL systems because the systems themselves produce this type of noise. NEXT can be reduced by reducing the frequency of operation because attenuation rises by about 5 dB per octave, while NEXT increases about 15 dB per octave. Crosstalk couplings are predominantly capacitive and so the coupled NEXT echo increases in level as frequency increases. It is the improvement in the signal-to-NEXT ratio that occurs at the lower frequencies of operation echo cancellation allows that makes HDSL feasible on the same two pairs that would otherwise require repeaters if one pair were used in each direction.

Although NEXT imposes a fundamental limit on performance, the limit in practice appears to be due to impulsive noise. Much of this is due to the traditional PSTN signaling used on adjacent pairs. In particular, dialing pulses, nonzero switched ringing, and relay bounce produce high-voltage transients that couple in capacitively to produce pulses of noise. The main energy content of these noise pulses is below about 40 to 100 kHz, which puts them more into the basic digital transmission band rather than the HDSL band. Unfortunately, when these pulses do corrupt HDSL transmission, they are likely to corrupt several adjacent bits because of their width. Error correction coding can be used to compensate for this effect of impulse noise, but unfortunately this introduces additional delay.

Inductive noise may also present a problem that is larger than consideration of frequency alone would indicate. This is because inductive noise can produce nonlinear effects, such as saturation of the coupling transformers, which may be used to interface the electronics to the line. Coupling transformers have so many advantages and are so small for high-frequency operation that they are likely to be used for HDSL for some time.

### 2.4.4 Standards for HDSL Transmission

The initial specifications for HDSL transmission in the United States called for dual duplex operation with a 2B1Q line code. The choice of line code is based on speed of implementation, since 2B1Q is the U.S. standard for basic digital transmission, and since other line codes could have had some technical advantages.

In Europe, 2B1Q has also been initially agreed on, but with two options. The first is a triple duplex option, which is not ideal because an additional pair is required, but which has the merit that it can take advantage of technology developed in the United States. The second option is a dual duplex approach operating at a higher rate. The goal for operating range in Europe is 4 km of 0.4-mm line.

It is possible that these standards will be modified in the light of the developing practical experience of HDSL systems, but once systems are deployed, operators will be reluctant to change because of the overhead costs that such a change will incur.

## 2.5 ADSL AND VDSL

The fundamental limit due to NEXT that occurs in HDSL transmission does not exist if the transmission is only in one direction, because then there is no received signal for the NEXT to interfere with. This is the principal factor that makes *asymmetrical digital subscriber line* (ADSL) technology feasible. In ADSL transmission, there is high-bit-rate transmission from the exchange to the customer and a lower-bit-rate transmission from the customer to the exchange, which is designed to be received in the presence of the NEXT from the higher downstream rate.

### 2.5.1 Objectives for ADSL Transmission

HDSL transmission should be able to support services required by business customers who are able to make use of bidirectional transmission at its high bit rate. ADSL transmission is seen as a complementary approach, which is more appropriate to residential customers.

The initial goal of ADSL was to support transmission at 1.544 or 2.048 Mbps downstream to the customer while supporting a lower rate control channel in the upstream direction on the same pair that also supplies a conventional PSTN or basic rate ISDN connection. The achievable line length has to be greater than for HDSL transmission, because residential customers often have longer lines than business customers. This may be overly ambitious because PSTN on the same pair produces excessive impulse noise, and because the low-frequency spectrum reserved for ISDN would be better used by combining the downstream ISDN transmission with the downstream ADSL transmission, since ADSL techniques would make better use of this spectrum and no guard band between the narrowband transmission in the baseband and the ADSL transmission would be needed (see Figure 2.15). Unfortunately, integrating the ADSL transmission and the narrowband transmission leads to operational problems, because local mains powering is likely to be needed for the ADSL transmission and this makes it difficult to maintain emergency narrowband service in the event of a mains failure.

The goal in the United States is to have a 1.544-Mbps downstream channel on most of the nonloaded pairs. About 25% to 30% of lines in the United States have loading coils, but these are typically the longer lines that would be difficult to use anyway. Operation is desired on single pairs consisting of up to

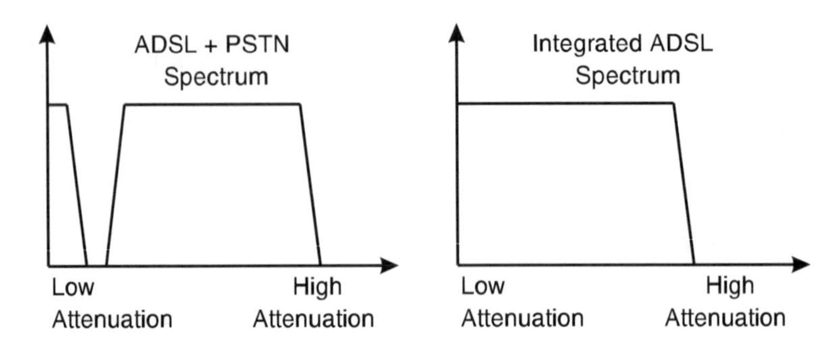


**Figure 2.15** ADSL spectrum allocations.

18 kft of 24-gauge wire, which is 50% further than HDSL can operate over its two pairs and is similar to the target for basic rate ISDN operation. The target *bit error rate* (BER) is $10^{-7}$. A margin of 6 dB or more above the theoretical signal-to-noise ratio is required to achieve this in order to take account of practical limits on implementation, such as DSP limitations and noise produced by the electronic circuits. There are about 75 million lines in the United States that could be used.

Equivalent operation is required in Europe, but the higher downstream rate of 2.048 Mbps limits the line length to less than that achievable at 1.544 Mbps, so it may not be possible in Europe to operate on a line that could support basic rate ISDN. On the other hand, operation in the United States can be more difficult in some cases because bridged taps are more common than in Europe, and equivalent range to that which can be achieved for basic rate ISDN is also difficult, even at the reduced rate of 1.544 Mbps.

In the longer term, it is hoped that operation at an even higher rate will be achieved, because this would allow a better quality of compressed video to be transmitted downstream.

### 2.5.2 ADSL Techniques

The greater range of ADSL transmission in comparison to HDSL is due to the absence of the high-speed return path from customers, because this eliminates the NEXT that would otherwise interfere with the reception of the high-speed signal at the customer's end. The absence of this NEXT creates a quieter environment, which allows the customer to receive the high-data-rate downstream signal.

Three main transmission techniques were considered for ADSL transmission: *quadrature amplitude modulation* (QAM), *carrierless amplitude/phase* (CAP) modulation, and *discrete multitone* (DMT) modulation. In QAM the source data stream is split into two streams at half the original source rate and these are used for discrete amplitude modulation of the sine and cosine phases of a carrier. The transmitted QAM signal is obtained by adding together the two modulated carrier phases (see Figure 2.16).

The CAP approach is similar to QAM, but avoids the use of a carrier. The source data stream is again split in half, but instead of the two halves being used to modulate two phases of a carrier, they are instead passed through two different filters and then added together. These filters form a Hilbert pair, which means they have the same amplitude response against frequency, but their phase responses differ by 90 deg. The result is similar to QAM because if the two half rate stream were each used to modulate the same carrier, then the filters would shift their phases by 90 deg, effectively producing a QAM signal. The difference for CAP is that the same shift in phases is still produced by the two filters in the absence of a carrier.

Whereas QAM uses a single carrier and CAP uses no carrier, DMT uses a number of carriers (see Figure 2.17), each of which supports a DMT subchannel that has a capacity determined mainly by its attenuation and noise characteristics. DMT transmission can be thought of as a number of QAM systems operating in parallel, one on each of the DMT carriers. The source data stream is divided up into symbol periods, each of which consists of a fixed number of bits, and these bits are distributed over the DMT subchannels. Bits are allocated to subchannels according to their capacity, and this can be varied adaptively to achieve approximately the same error rate on each. The DMT approach was eventually ratified by ANSI as the standard modulation scheme for ADSL in 1995.

There are a number of simplifications that occur when a DMT transmission system is implemented. There is no great need for equalization to compensate for pulse dispersion on the DMT subchannels, because these have

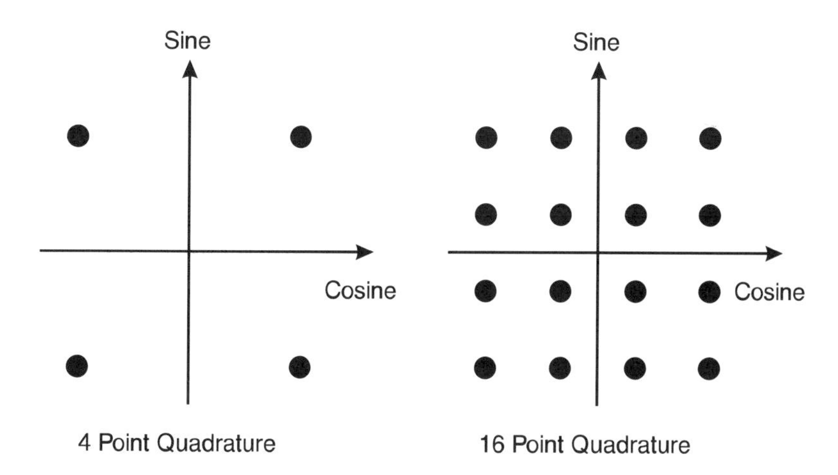


**Figure 2.16** A 4- and 16-point QAM.

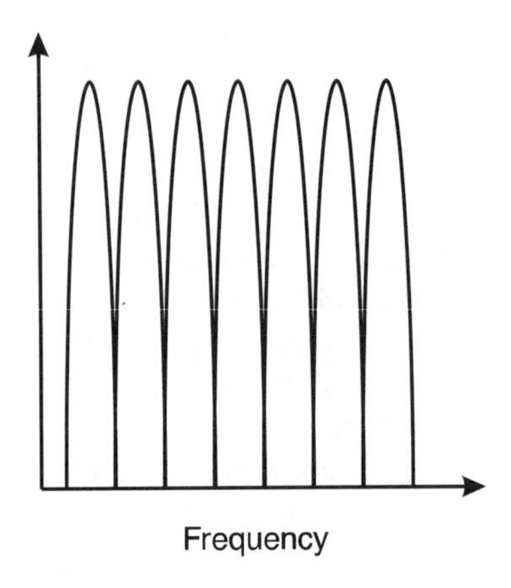


**Figure 2.17** DMT spectrum.

approximately linear transmission characteristics, since the bandwidth of each channel is narrow. The subchannels must be chosen to be orthogonal over the symbol period, and this is consistent with DSP using a fast Fourier transform.

DMT also has inbuilt immunity to impulsive noise, because the energy of the noise is spread across the large number of subchannels, and this can be enhanced by forward error correction with bit interleaving to spread errors over a number of blocks. The form of DMT that has been adopted by ANSI gives protection over three symbol periods, enabling it to survive impulses 500 μs long.

The QAM and CAP approaches have similar levels of complexity and achieve similar levels of performance. Spectral compatibility between different copper pair transmission systems is facilitated by the DMT approach because it has the flexibility to dynamically adjust its power spectral density to compensate for interworking with other systems. This advantage is greater for shorter loops because the greater number of usable subchannels give greater flexibility. The flexibility of DMT also allows for reconfiguration of the upstream and downstream data rates.

Unfortunately, there is a price to pay for the flexibility of DMT. DMT implementations have a higher signal processing delay, which is less of a disadvantage on shorter loops if the data rates are higher and symbol periods shorter. In addition, the echo cancellation required for DMT is greater because it must be performed on each subchannel.

### 2.5.3 ADSL Services

The services that can be supported by ADSL systems are different from those intended for HDSL systems because ADSL is targeted at residential users. The characteristic feature of ADSL services is that they provide a lot of information to the customer, but only have a small amount of information sent by the customer.

The video compression techniques standardized by the Moving Picture Experts Group (MPEG) allows video of a quality comparable to that provided by a video cassette recorder to be stored on a CD-ROM. This form of compressed video is also suitable for transmission downstream over an ADSL link, because the access speed of a CD-ROM is less than the ADSL transmission rate. The combination of MPEG and ADSL technology enables a number of different services to be provided. Higher rates of ADSL operation would allow even better quality of video, such as that using MPEG-II coding, to be transmitted.

The most widely publicized service of this type is probably *video-on-demand* (VoD), which allows a user to select a video for viewing from a video library. Home shopping, which allows a user to browse in a virtual supermarket, is another service of this type that has been well publicized. It has also been suggested that this technology could be used for teaching because it allows the teacher or lecturer to present the material and to receive and answer questions during the presentation. This would be of particular use where a school was too small to have a specialized teacher in a particular subject or where a lecture is given by some special expert. Politicians could also use this technology to interact with a larger number of their supporters in a virtual rally, so it may be wise to remember that any technology can be misused.

There are also a number of services that do not necessarily require the MPEG coding of video signals. These can be based on the ease of access to a CD-ROM library and have the advantage that it is not necessary to buy or be in physical possession of the CD-ROM to use the service. For specialized applications, it may be more economical to access a remote database rather than acquire and maintain an individual installation. In other applications, it may be desired to sample the information, for example, through CD-quality audio-on-demand, before buying or downloading it. There may also be a demand for services based on interactive graphics, such as highly detailed, multiplayer interactive games.

In addition to these primary services, there may be secondary services such as intelligent advertising, which is not necessarily the contradiction that it appears to be. Services that are supported by advertising could have the advertising intelligently tailored to particular user groups. Alternatively, the suppliers of the services could sell information on the statistical characteristics of their users to third parties who could use it to define and target their products.

However, the service that is possibly the most dynamic driver for ADSL has not arisen as a planned telecommunications service, but instead has arisen from the needs of computer communications. This service is Internet access. The asymmetrical data rates that characterize Internet access make it especially well suited to ADSL, and for many people, fast interactive access to the Internet has a higher priority than a better video service.

### 2.5.4 Limits to ADSL Operation

Although ADSL systems are not restricted by NEXT in the same way as HDSL systems, they are intended to operate on longer lines to support residential services and at higher rates because they are restricted to a single copper pair. This makes them highly susceptible to impulse and other types of noise.

ADSL systems are affected by the same types of impulse noise that affect HDSL and DSL systems. They may have the advantage that there is a lower probability of noise being generated by heavy electrical machinery in a residence, but this may be offset by the large number of electrical appliances. There are also sources of radio frequency interference at residential sites, which may change as technology evolves. Trials have now been carried out with error correction coding to provide compensation for the effects of noise.

The greatest source of interference with ADSL transmission may not be external, but may be the interference produced by *plain old telephone service* (POTS) signaling. POTS signaling on adjacent pairs is one of the main sources of impulse noise for HDSL systems. For ADSL operation, POTS signaling shares the same metallic pair. Ring trip in particular can produce large transient voltages. Although these transients can be filtered out, they can still swamp the received signal through leakage currents and nonlinear effects. Even if the interference due to POTS on the same pair is removed by good filter design, interference from adjacent pairs will still remain.

Although ADSL systems are not limited by the NEXT they themselves produce, they may be susceptible to the NEXT from other systems that transmit from the customers' ends. This can be alleviated by staggering the frequencies of operation of DSL and HDSL systems that share the same cable. Care must be taken here to avoid the lobes in the spectrum for ISDN basic rate transmission that could also interfere with ADSL operation. The NEXT produced by the remote ends of HDSL systems may be even more significant. It has been suggested that HDSL systems will not cause problems because they operate on lines that are shorter than those used for ADSL. This appears to be wishful thinking, since ADSL systems will operate on all lengths of lines. It may also be a problem when HDSL and ADSL are used on the final drop to a customer at the end of an optical fiber system over the same cable.

Guidelines are required for the combined deployment of HDSL and ADSL systems because of the NEXT generated by HDSL systems. It may also be necessary to have some form of automatic modeling of the cable utilization because manual records of cabling are notoriously unreliable.

### 2.5.5 Very-High-Speed Digital Subscriber Loop

VDSL transmission may be used at the end of an optical fiber link for the final drop to the customer over a copper pair. In *fiber-to-the-curb* (FTTC) systems, the VDSL tail may be up to 500m long, and rates of 25 to 51 Mbps are being considered. In *fiber-to-the-cabinet* (FTTCab) systems, the tail may be over a kilometer, and rates of 25 Mbps are being considered.

The higher data rates for VDSL make DMT an attractive approach, especially because of its adoption for ADSL by ANSI. However, it may be better to use different channels for different directions of transmission, because this is easy to implement on a multichannel system, especially if the data rates are asynchronous. As for ADSL, the performance of the DMT for VDSL can be improved by bit interleaving and forward error correction.

The spectrum for VDSL transmission is unlikely to extend much beyond 10 MHz for practical systems, as compared to about 1 MHz for ADSL transmission. However, it may start at a higher frequency of about 1 MHz to reduce the interaction with other transmission systems at lower frequencies and to simplify the filter specification. Power levels for VDSL need to be lower than for ADSL because copper pairs radiate more at higher frequencies, generating greater electromagnetic interference.

## 2.6 SUMMARY

Copper pairs have been the traditional medium for access in telecommunications networks. Aluminum has been tried, but was found to have problems with jointing. Concerns about increasing relative cost of copper were not justified, but aluminum pairs are still present in some networks.

High-penetration digital transmission over copper pairs was developed to support ISDN basic rate access. The de facto international standard for this is the ANSI U-interface standard, which was developed in the United States to avoid the ownership of equipment by operators at a customer's premises. This standard determines the coding and transmission techniques used and ensures that digital transmission for basic rate ISDN is possible on all but a small minority of existing pairs.

Analog pair-gain systems have been used for many years to permit two users to share the same copper pair. More recently, digital pair-gain systems have been developed based on the digital copper pair transmission systems used for basic rate ISDN and on digital speech compression techniques. These allow up to eight users to share the same copper pair.

Higher speed digital transmission, up to a few megabits per second, can be used on a large percentage of existing copper pairs, because the majority of copper pairs are relatively short. Copper pairs can also be used in sophisticated asymmetric transmission systems where a relatively low-speed upstream link is used to control a higher rate downstream service, possibly even on a different medium. Higher rate transmission can also be used when a copper pair is used as the final drop following optical fiber link.

All sophisticated high-speed copper pair systems, from analog pair-gain systems onwards, operate better on "clean" copper pairs (i.e., without loading coils, which kill high-speed transmission, or bridged taps, which can degrade performance and may require the use of decision feedback equalizers). They also have problems with powering, since there is often difficulty in powering more than a conventional telephone from a single copper pair.

## Selected Bibliography

Baker, G., "High Bit-Rate Digital Subscriber Lines," *Electronics & Communication Engineering J.*, Vol. 5, No. 5, October 1993, pp. 279–283.

Barton, M., and M. L. Honig, "Optimisation of Discrete Multitone to Maintain Spectrum Compatibility With Other Transmission Systems on Twisted Copper Pairs," *IEEE J. Selected Areas in Communications*, Vol. 13, No. 9, December 1995, pp. 1558–1563.

Collins, R., "Planning the Irish Rural Network," *International Conf. on Rural Telecommunications*, May 1988, pp. 114–119.

Ferland, P., A. Sticca, and G. Koleyni, "ISDN Trials on Subscriber Loops and Customer Premise Wiring," *ISSLS 88 Proc.*, pp. 47–51.

Gildenhuys, B. J. F., "The Design of an N*64k Data Transmission System Based on Multiple Parallel ISDN U-Interfaces," *COMSIG'92*, pp. 205–210.

Giradeau, J., et al., "ISDN U Transceiver Algorithm, Development System, and Performance," *GLOBECOM '89 Proc.*, Vol. 3, pp. 1957–1965.

Henkel, W., T. Kessler, and H. Y. Chung, "Coded 64-CAP ADSL in an Impulse-Noise Environment—Modelling of Impulse Noise and First Simulation Results," *IEEE J. Selected Areas in Communications*, Vol. 13, No. 9, December 1995, pp. 1611–1621.

Hsing, T. R., C.-T. Chen, and J. A. Bellisio, "Video Communications and Services in the Copper Loop," *IEEE Communications Magazine*, Vol. 31, No. 1, January 1993, pp. 62–68.

Kerpez, K. J., "Forward Error Correction for Asymmetrical Digital Subscriber Lines (ADSL)," *GLOBECOM '91*, Vol. 3, pp. 1974–1978.

Kerpez, K. J., "The Range of Baseband ADSLs as a Function of Bit Rate," *GLOBECOM '92*, Vol. 1, pp. 40–44.

Lechleider, J. W., "High Bit Rate Subscriber Lines: A Review of HDSL Progress," *IEEE J. Selected Areas in Communications*, Vol. 9, No. 6, August 1991, pp. 769–784.

Piasetsky, J., D. Zinger, and M. Shalit, "Smoothing the Transition From Copper to Fibre," *IEE Colloquium on Customer Access—the Last 1.6 km*, June 1993, pp. 3/1–7.

Riley, G. I., and G. J. Harris, "Pair Gain," *IEE Colloquium on New Techniques in Providing Customer Services With Copper*, December 1989, pp. 4/1–3.

Szechenyi, K., and K. Bohm, "Impulsive Noise Limited Transmission Performance of ISDN Subscriber Loops," *ISSLS 88 Proc.*, pp. 29–34.

Takahashi, Y., et al., "An ISDN Echo-Cancelling Transceiver Chip Set for 2B1Q Coded U-Interface," *IEEE J. Solid-State Circuits*, Vol. 24, No. 6, December 1989, pp. 1598–1604.

Walkoe, W., and T. J. J., Starr, "High Bit Rate Digital Subscriber Line: A Copper Bridge to the Network of the Future," *IEEE J. Selected Areas in Communications*, Vol. 9, No. 6, August 1991, pp. 765–768.

Waring, D. L., "The Asymmetrical Digital Subscriber Line (ADSL): A New Transport Technology for Delivering Wideband Capabilities to the Residence," *GLOBECOM'91*, Vol. 3, pp. 1979–1986.

Waring, D. L., J. W. Lechleider, and T. R. Hsing, "Digital Subscriber Line Technology Facilitates a Graceful Transition From Copper to Fiber," *IEEE Communications Magazine*, Vol. 29, No. 3, March 1991, pp. 96–104.

Werner, J.-J., "The HDSL Environment," *IEEE J. Selected Areas in Communications*, Vol. 9, No. 6, August 1991, pp. 785–800.

Young, G., and N. G. Cole, "Design Issues for Early High Bit-Rate Digital Subscriber Lines," *GLOBECOM'90*, Vol. 2, pp. 1177–1182.

Young, G., K. T. Foster, and J. W. Cook, "Broadband Multimedia Delivery Over Copper," *BT Technology J.*, Vol. 13, No. 4, October 1995, pp. 78–96.

Zogakis, T. N., J. T. Aslanis, and J. M. Cioffi, "A Coded and Shaped Discrete Multitone System," *IEEE Trans. Communications*, Vol. 43, No. 12, December 1995, pp. 2941–2949.

# 3 Optical-Fiber Technology

> "Lighten our darkness, we beseech thee."
>
> —*The Common Prayer Book*

## 3.1 BACKGROUND

Optical fiber offers a number of advantages over coaxial cable for transmission in the core network, notably with respect to bandwidth, attenuation, and cost. In particular, the need for repeaters to compensate for signal attenuation on high-traffic coaxial links in the core network can be reduced if optical-fiber links are used, because the attenuation of optical fiber is much less than that of coaxial cable. This eliminates the maintenance costs, powering difficulties, and reliability hazards associated with large numbers of repeaters. For submarine links, especially intercontinental links across oceans, these advantages have been overwhelming.

The cost reductions possible through the use of optical fiber in the core network has encouraged the development of optical-fiber technology and has led to suggestions as to how optical fiber can be used in access networks. This is not as simple as in the core network because there are no immediate cost savings. The story of optical access networks is really the story of the search for cost-effective approaches to the use of optical fiber in the access network and for revenue generating high-bandwidth services to justify it.

*Synchronous digital hierarchy* (SDH) technology will not be covered here. This is not because SDH has no role to play in access networks, but because SDH links and rings are more likely to be used as feeder transmission systems to support other forms of access technology. SDH may become more directly relevant when customers require ATM interfaces, because SDH is one of the options for the ATM physical layer, but initially for access networks SDH is more of a supporting technology than a primary transmission technology.

## 3.2 OPTICAL AND OPTOELECTRONIC COMPONENTS

One of the challenges for optical access networks is to reduce the cost of the components necessary for their implementation. The most obvious components are the fiber and the associated laser transmitters and photodiode receivers. For passive multistar architectures, it is also necessary to have splitters and possibly directional couplers. Optical filters are required in front of the photodiodes to limit the optical wavelength received by them if several optical wavelengths are used simultaneously. Upgrading to broadband operation typically requires both optical filters and, more significantly, optical amplifiers.

### 3.2.1 Optical Fiber

The light that travels down an optical fiber is confined within the fiber by total internal reflection at the boundaries. A light ray incident on the boundary between a medium with a relatively high refractive index, such as glass, and a medium with a relatively low refractive index, such as air, is completely reflected back into the high refractive index medium if the angle of incidence is sufficiently shallow. The light does in fact tunnel a slight distance into the low-index material before being completely returned (see Figure 3.1). This phenomenon corresponds to quantum tunneling and can be used in splitters and couplers, since it allows photons to tunnel between closely adjacent fibers.

For monomode optical fiber, the diameter of the active core of the fiber is about the same size as the wavelength of the infrared light it carries. If the diameter is about twice this size or greater, then higher order transverse modes can occur in the transmission, where the transverse waveform exhibits appropriate standing waves similar to those on a skipping rope (see Figure 3.2). These transverse modes spread the optical pulses as the different modes have different transmission times, since they correspond to different lengths of transmission paths.

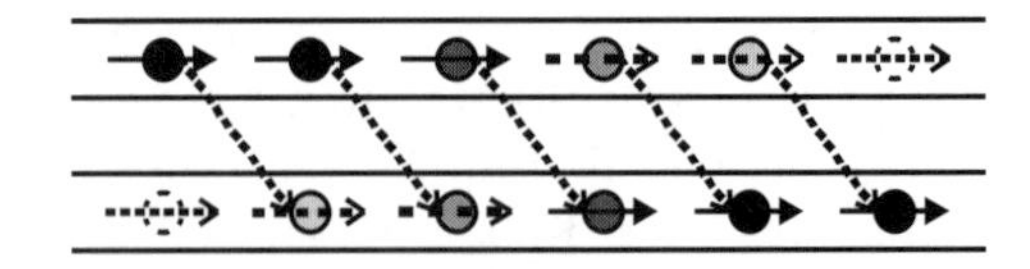

**Figure 3.1** Tunneling at the boundary of a fiber.

**Figure 3.2** Transmission modes within an optical fiber.

Currently the most popular type of optical fiber has a sharp change in refractive index between the core and the cladding, and the cladding is sufficiently thick to prevent the escape of the internally reflected signal by tunneling. Earlier types of optical fiber, known as *graded fiber*, had a gradual change of index and a core that was larger than the wavelength of the light transmitted. The grading in the refractive index reduced the optical path length for signals that bounced from side to side and acted to equalize the transmission times of the different transmission modes because the light traveled faster when it was at the sides. As fabrication techniques improved, graded fiber became less popular because it was easier to produce step-index monomode fiber, which does not exhibit this type of spreading of the optical pulses.

### 3.2.2 Optical Transmitters

It is now well accepted that laser diodes are the best choice for the transmitters in optical access networks. Simpler *light-emitting diodes* (LED) are not as effective, because their wider optical spectrum limits the utilization of the fiber, since it prevents the simultaneous use of a large number of different optical wavelengths as different optical channels. LEDs also create larger pulse dispersion, since the different wavelength components arrive at the far end at different times because chromatic dispersion of the fiber produces different transmission times for different wavelengths.

The cost of laser diodes is mostly determined by the size of the market for them. The physical processes used to produce a laser diode for optical access networks are almost identical to those used for *compact disc* (CD) players. The cost of these two types of diodes is very different because there is a large market for CD applications, while the market for telecommunications applications is in its infancy. The cost of the transmitters and receivers for optical telecommunications is a reflection of the fact that the market only supports a high-tech "cottage industry."

Much attention has been focused on ways to increase the size of the telecommunications market. It has even been suggested that CD lasers at about 0.78 μm should be used because they are so inexpensive. This suggestion is not feasible, because this wavelength is not well suited to the loss and dispersion characteristics of the optical fibers. The differential or group dispersion of optical fiber has a natural minimum at about 1.3 μm, and operation at this wavelength minimizes the spread of optical pulses. The attenuation of the fiber has a minimum at about 1.5 μm, and the group dispersion can be shifted to this wavelength by the use of boundary effects to produce dispersion-shifted fiber.

Another approach to reducing the cost of components through increasing the market size has been to seek agreement on international standards for laser diodes. This would encourage cost reductions through the creation of a common international market. It may be even more effective to combine the telecommunications market with the computer communications market for the *fiber-optic data distribution interface* (FDDI) optical transmission standard at 125 Mbps. This would allow the telecommunications components to take advantage of a more mature market for computer communications, although higher data rates would be needed for broadband operation.

### 3.2.3 Optical Coupling and Optical Splitters

Light confined to an optical fiber by total internal reflection is still sensitive to conditions outside the core of the fiber. Although no transmitted signal leaves the core of the fiber, there are exponentially decaying evanescent electromagnetic fields outside of the fiber. This phenomenon, which is an optical quantum effect and which can be described in terms of wave mechanics, can be used in the construction of optical couplers and splitters.

If two optical fibers are brought close together, then the evanescent fields from the light in one fiber react to the presence of the second fiber and allow light to couple into it. If light is thought of in terms of particles, then another way of expressing this is to say that the photons from the first fiber can tunnel into the second fiber if it is sufficiently close by. If the coupling between the two fibers is sufficiently strong, then the light that was originally traveling in one fiber is distributed evenly between both fibers. This arrangement creates a 1:2 optical splitter (see Figure 3.3). Splitters with a larger split ratio can be created by cascading these simple splitters or by coupling between multiple fibers. These higher ratio splitters can be used to create a passive multistar topology. The loss of optical signal through splitting is the dominant loss in the downstream transmission path. Technically, it is not a loss at all, because the signal that is "lost" is the signal that is transmitted to the other remote ends.

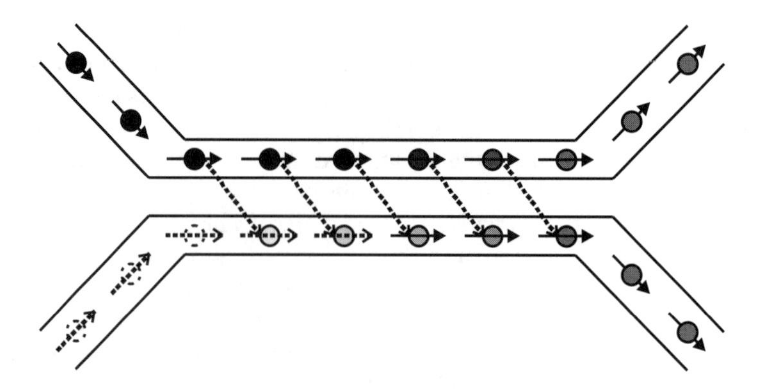

**Figure 3.3** Optical coupling and optical splitting.

Unfortunately, optical splitters also create loss in the upstream direction because they operate equally well in both directions. Light traveling upstream through a simple 2:1 splitter from one leg couples into the adjacent fiber. The equal distribution of the light between the two coupled fibers causes half of the light transmitted upstream to miss the correct upstream leg and get lost in the dummy tail of the coupler.

A 2:1 splitter can also be used in reverse to allow a detector to receive the signal from the remote end without being swamped by the signal being transmitted locally. The locally transmitted signal does not reach the receiver because it is traveling in the wrong direction, but this means that half of the transmitted signal is lost in the dummy tail of the coupler. This application is known as a *directional coupler*. Unfortunately, the price for this approach is rather high, since half of the received signal misses the detector and half of the transmitted signal is lost in the dummy tail.

If directional couplers are used at both ends of the link, then three-quarters of the photons traveling in either direction are lost. This inefficiency appears to find a fundamental resonance in the depths of human perversity, because there has been a prolonged reluctance to consider technically superior alternatives to achieve duplex operation, such as ping-pong techniques with integrated transmitter and receiver diodes.

### 3.2.4 Optical Filters

There is an asymmetry between the optical bandwidths of operation for laser transmitters and for photodiode receivers. Laser diodes transmit in a narrow spectral range, while photodiodes are sensitive across a wide spectral range. If several optical wavelengths are to be used simultaneously, then it is necessary to limit the sensitivity of the photodiode receivers by interposing an optical filter between them and the optical fiber.

Optical filters typically operate by making use of constructive and destructive interference between adjacent boundaries (see Figure 3.4). If successive

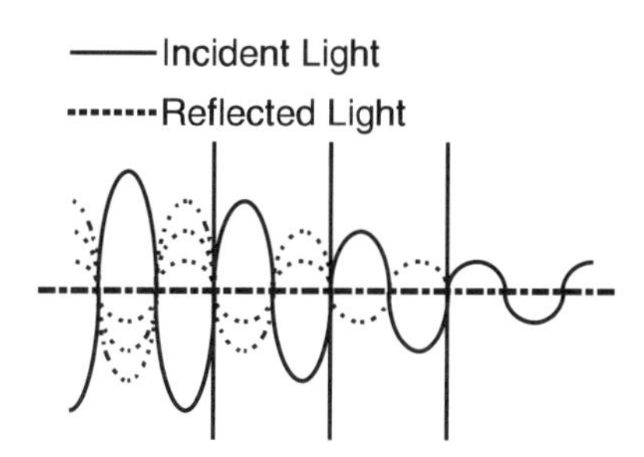


**Figure 3.4** Filtering from constructive interference of reflected waves.

layers of high and low refractive index materials have the correct thicknesses, then the combination of the phase shift through the layer and the phase shift due to reflection at the boundaries results in the near total cancellation of the reflected wave at a particular frequency. By tailoring the thicknesses of a large number of layers of different refractive index, the optical characteristics can be adjusted to allow transmission at only the desired optical band and reflection at other wavelengths.

Optical filters are essential to allow upgrade to broadband operation, because they allow new broadband transmission systems to be added to the fiber infrastructure with minimum disruption to existing narrowband systems. The filters incorporated at the narrowband receivers prevent those receivers from responding to the optical frequencies used for the broadband systems. If these filters are not included in the narrowband systems at the outset, then they will have to be retrofitted before the network is upgraded to broadband operation.

### 3.2.5 Optical Receivers

Photodiode receivers operate through the creation of hole-electron pairs in a reverse-biased p-n semiconductor junction when a photon is absorbed. These receivers are sensitive to a wide range of optical frequencies, because any photons that have sufficient energy to create hole-electron pairs generate a photocurrent in the junction. Even in the absence of incident photons, a dark current will flow at room temperature due to the migration of minority carriers across the p-n junction.

The efficiency of the photodiode receiver can be characterized by the number of hole-electron pairs created, on average, by an incident photon. On an ideal receiver, each photon would produce a pair and the efficiency would be 1. Practical limitations reduce the efficiency, since photons can be scattered or otherwise lost without producing a pair.

It is tempting to think of each pair produced as corresponding to a binary digit, but this is misleading because it does not take account of coding and noise. Coding allows the possibility of several bits to be encoded per pair, since, for example, the time interval between pulses could be used to represent binary words. Noise has the opposite effect, since several pairs may be needed to give a signal that can be reliably detected against the noise background. It is more correct to treat each pair as part of a current pulse and to analyze performance in terms of the number of pairs per pulse. The coding used on the pulses and the magnitude of the pulses required to overcome systems noise are then separate issues.

There can be practical advantages to using commercially available receivers designed to operate at higher rates and using these with narrow, high-amplitude transmitted pulses. Using commercially available receivers reduces the cost, since these have an established volume market. Narrow, high-amplitude pulses keep the optical power down while increasing the electrical power in the receiver. This is because electrical power is proportional to the square of the optical power, since the electrical current is proportional to the optical power. This higher electrical power gives a better electrical signal-to-noise ratio. This approach can also be combined with the position of the pulse to increase the information transmitted.

The mechanism of hole-electron pair production limits the potential of optical-fiber systems, because it translates considerations of coding and noise into requirements of photons per bit. It is possible to beat the theoretical limit of one pair per photon if optical amplifiers are used, since a single photon can stimulate the emission of several photons in the optical amplifier, which in turn produce several hole-electron pairs. Several hole-electron pairs can also be produced by an avalanche photodiode, because the first pair produced gains energy from an applied electric field and then additional carriers are produced by energetic collisions with atoms. However, the sensitivity of avalanche photodiodes is compromised by secondary effects, such as spontaneous pair creation caused by the high voltages needed for their operation.

For operation at the wavelengths needed for optical access networks, the size of the depletion region between the p-type carrier and n-type carrier doped regions of the semiconductor needs to be increased to improve the absorption of the longer wavelength photons. This is achieved by reducing the doping to create what is effectively an intrinsic level of doping between the two highly doped regions, to produce a PIN photodiode. Extremely sensitive operation can then be achieved if a PIN photodiode is combined with an optical preamplifier.

### 3.2.6 Dual-Mode Diodes

There are significant advantages to the use of the same semiconductor diode as a laser diode for transmission and as a photodiode for reception. The use of a single component in this way, however, has the limitation of being best suited to systems using a ping-pong transmission technique, since it cannot operate in both modes simultaneously. It also has the disadvantage of being a compromise design with less than optimal performance in either mode. It is not reasonable to expect that a semiconductor diode that is designed to operate both as a laser and as a photodiode will have the same performance in either mode as diodes that have been optimized to operate only as lasers or as photodiodes. The key issue is whether the disadvantages outweigh the advantages of improved power budget and reduced cost.

A dual-mode diode can result in a system that has an improved power budget, since it allows the elimination of optical couplers at either end of the optical fiber. These couplers are used to separate the direction of transmission at either end of the fiber so that the receiver is not swamped by the transmitted signal. The net result is that it is possible for a dual-mode diode to give a better power budget for systems and have the lower operational costs of single-fiber working. A dual-mode diode must also operate at a higher speed in each direction because time is shared between the two modes. This is not necessarily a disadvantage, since there are signal-to-noise advantages to using narrower, higher amplitude pulses. Optical fiber is also a good medium for this coding technique because it does not restrict the bandwidth in the same way as copper pairs do.

Another advantage of dual-mode diodes is that only half the number of diodes are needed in the system, since each dual-mode diode replaces two conventional diodes. How great an advantage this is depends on the markets for the different types of diodes, since the actual costs depend on the level of automation in the production the market invites. From a quantitative perspective, however, it is reasonable to expect system costs to be lower if one component can be used instead of two.

It is not reasonable to use a dual-mode diode when simultaneous operation in both directions is required, because in this case two diodes are required, so diodes can be used that are optimized for operation in each mode. For ping-pong operation, however, a dual-mode diode is a natural choice for new systems.

### 3.2.7 Optical Amplifiers

Optical amplifiers have had a revolutionary impact on optical access network concepts because they facilitate the upgrade to broadband operation and because they offer a simpler and more robust alternative to optical heterodyne techniques.

A fiber-optic infrastructure that has been optimized for narrowband telecommunications services presents difficulties for broadband upgrade because of the optical power budget. For the narrowband systems, the optical power that is output at the headend is distributed to all of the remote ends to provide adequate photons per bit for the lower data rates of the narrowband services. To maintain a sufficient number of photons per bit for broadband services, the transmitted optical power would have to be increased because the bit periods are smaller. Raising the optical power levels to a suitably high level can cause operational problems because of safety requirements and may not be feasible in any case because of the limits to output power on laser diodes. Optical amplifiers can be used to avoid this problem because they can be collocated with optical splitters to offset the distribution losses at the splitters so that there are no high power levels on the fiber (see Figure 3.5.) They can also be used to boost the transmitted power of the lasers.

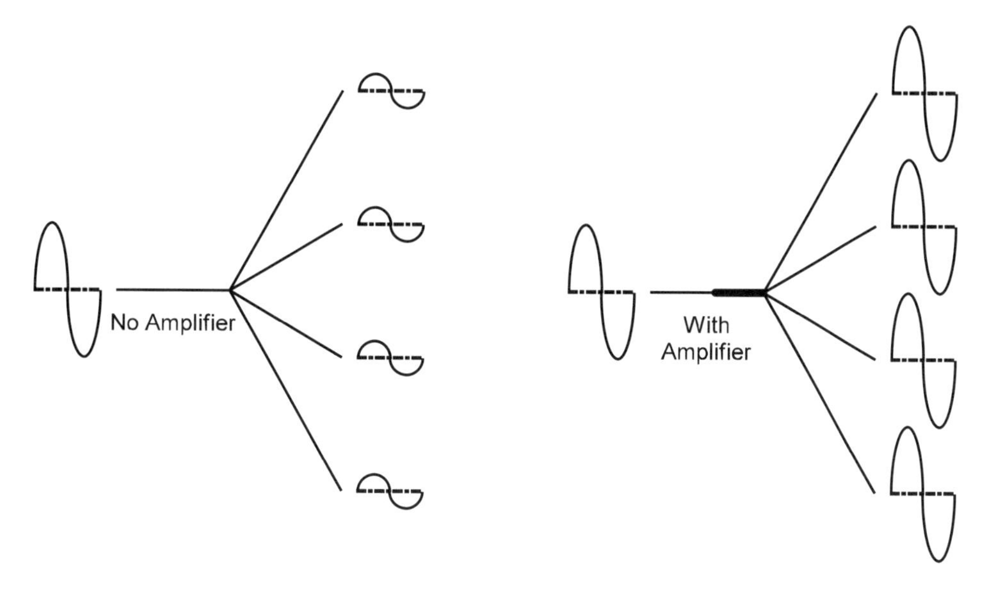


**Figure 3.5** Use of optical amplifier before fiber split.

Optical amplifiers are very simple structures consisting of a length of specially doped optical fiber. The doping atoms, typically rare earths such as erbium, are excited by the light of one particular wavelength and triggered into emitting photons through resonance with photons of the same wavelength belonging to the signal to be amplified. This process is almost identical to that used in a laser, but the spontaneous emission of photons by the excited atoms here is undesirable.

The physical construction of amplifiers is simplified because there is no conversion of light into an electronic signal. This makes for a simple, robust, low-cost amplifier, which can operate over a range of optical wavelengths. The original forms of optical amplifiers were designed to operate in the 1.5-μm window, which is especially useful, since this allows high-bit-rate signals within this window to be amplified while lower-bit-rate signals within the 1.3-μm window can bypass the amplifier without boosting. Amplifiers for the 1.3-μm window are also now available.

Optical amplifiers can also be integrated with photodiode receivers to increase the sensitivity of the receivers. Optical heterodyne and homodyne techniques were investigated in the years before the development of suitable optical amplifiers because they improve the sensitivity of the diode receivers. These techniques are the optical equivalents of the heterodyne and homodyne techniques used to improve sensitivity in radio communications, and operate through the positive and negative interference of the signal to be detected with a local signal of similar (heterodyne) or identical (homodyne) wavelength. The interference behaves like a modulation of the local signal and gives a better signal-to-noise ratio because on detection the received signal is multiplied by the local signal rather than by itself, which is weaker.

In practice, optical heterodyne and homodyne technologies suffer from a lack of suitably robust components, particularly local laser oscillators, which are both stable and tunable. Optical amplifiers are simpler and more robust. A sophisticated local oscillator is not required, only a simple pumping signal to excite the doping atoms. An incoming photon resonates with the excited atoms, which coherently emit identical photons without the absorption of the incident photon, resulting in the creation of a number of hole-electron pairs when the coherent burst eventually reaches the photodiode.

### 3.2.8 Reflective Modulators

Monolithic, integrated reflective modulators that are electronically controlled have been used as a low-cost and low-power alternative to laser diodes at the remote ends of an optical access network. These devices operate by using an electrical voltage to alter the optical path length in an optical cavity, altering the phase relationship between different paths and so causing a change between constructive and destructive interference on reflection.

The major disadvantage of reflective modulators is that they rely on the reflection of an attenuated signal from the headend. This signal is attenuated by splitters on both legs of its round-trip journey from headend to remote end and back again. This double attenuation limits the degree of splitting that can be tolerated in the system.

On the positive side, there is a significant reduction in the number of lasers needed in the system. For example, an eight-way split would require two lasers if reflective modulators were used, since the laser that is used to transmit at the headend does not typically provide a suitable signal for use by the remote modulators, since it is already modulated. For an equivalent passive star or multistar topology, a total of nine lasers would be needed, since a laser would be needed at the headend and at each of the remote ends. Point-to-point operation on eight separate links would require 16 lasers. Using reflective modulators also has the advantage that the environment in the headend may be more benign than that at the remote end, since the headend is more likely to be in a protected environment.

As in the use of dual-mode lasers, the actual savings produced by using modulating reflectors depends on the relative cost of the components. In contrast, however, the application of modulating reflectors is likely to be specialized because only lower fiber split ratios can be supported, although these lower split ratios could simplify the upgrade to broadband operation using more conventional techniques.

## 3.3 FIBER ARCHITECTURES

In the search for a cost-effective approach to the deployment of optical fiber in access networks, the physical architecture plays a significant role. Costs are strongly affected by the topological configuration of the optical fiber and by the configuration used to provide service to individual customers at the remote ends. Consideration must also be given to the number of fibers per customer, the operating wavelengths on the fiber, and the ability to upgrade architectures from narrowband to broadband operation.

### 3.3.1 Fiber Configurations

Over the years, various configurations for the topology of optical access networks have been considered. Perhaps the simplest configuration is a single star (see Figure 3.6), which consists of individual fibers that radiate outwards from their common *optical line termination* (OLT) at the exchange with point-to-point optical transmission on each fiber to the *optical network units* (ONU) at the customer ends. This approach was used in some early access networks with a remote multiplexer connected at the ONUs to provide service to business users.

Unfortunately, the single-star configuration is only cost-effective to large business sites because of the high overhead cost of the optoelectronics required at each end of each fiber. The single-star topology is not effective for the more numerous smaller business or residential sites, since as the number of telephone lines served at each remote end decreases, the cost per line of the optoelectronics increases. In addition, with a large number of smaller sites, the cost of the cabling is larger because more optical-fiber cables are needed for the same number of telephone lines.

As the size of the remote sites becomes smaller, ultimately one telephone line is served by each fiber in a single-star configuration. At this point, optical

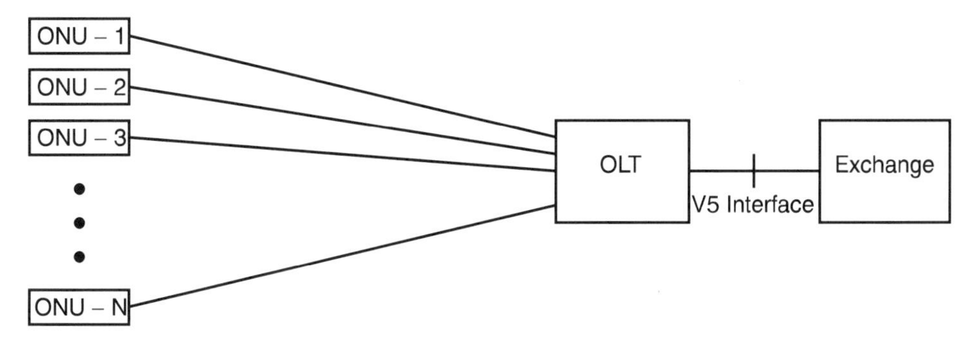


**Figure 3.6** Single-star fiber topology.

fiber is not cost-effective in comparison with traditional copper for narrowband services, because, although the cabling costs are comparable to those for copper, fiber also incurs the additional costs of the optical transmission system. Even the new services that could be provided on optical fiber might not be sufficient to make fiber more cost-effective, because many of these services could be provided on the existing copper network, using more sophisticated copper transmission techniques over existing copper pairs and so avoiding the expense of installing new cables for fiber.

The cost of fiber cabling is reduced if an active multistar architecture is used (see Figure 3.7). This allows a single fiber at the exchange to be shared between a number of remote fiber terminations. The active *multiplexer* (MUX) nodes contain electronics to perform multiplexing and optoelectronic devices at the inputs and outputs. They allow traffic from the different terminations to be multiplexed onto a common shared fiber. Unfortunately, although the cost of the cabling has been reduced, the cost of the transmission system and the installation costs have both been increased because of the presence of the active MUX nodes. The increase in these one-off costs, however, is overshadowed by the even higher recurrent costs of maintaining and powering the active nodes. In addition to these economic disadvantages, active multistar architectures have a technical disadvantage because they cannot be easily upgraded to take advantage of the transmission capacity of the fiber. They lack optical transparency, since the MUX nodes create an electronic bottleneck between the exchange and the customer, which prevents new higher rate services from being added without also upgrading the MUX nodes.

The critical innovation that has led to optical fiber being perceived as a viable transmission medium for access networks is the development of the passive multistar topology (see Figure 3.8), which has a similar advantage with respect to cabling costs as the active multistar, but has no active nodes to create the powering and maintenance problems and to prevent optical transparency. The passive multistar further reduces the cost of the optical transmission system, because a single optical transmitter and receiver at the exchange end

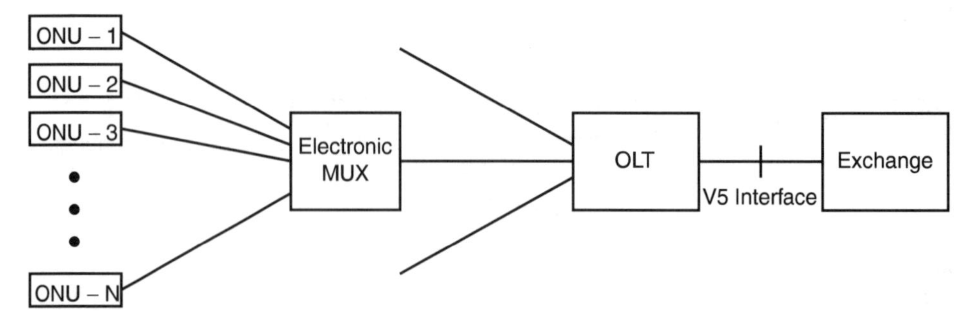


**Figure 3.7** Active multistar fiber topology.

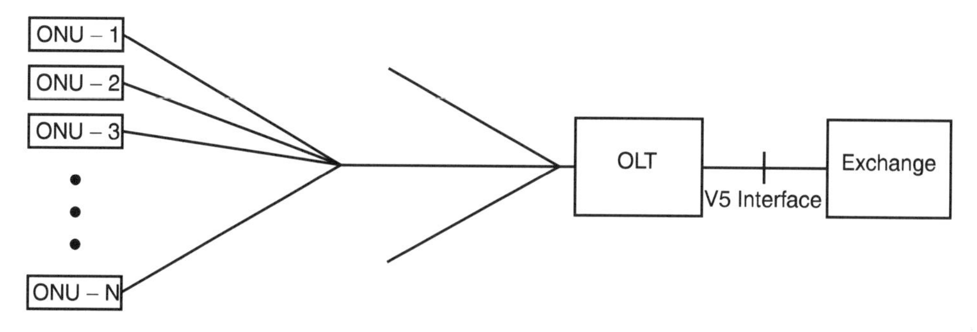


**Figure 3.8** Passive multistar fiber topology.

supports all the remote ends, approximately halving the high cost of the optoelectronics.

In a passive multistar configuration, the active MUX nodes of the active multistar configuration are replaced with passive splitters that distribute the optical signals from the OLT to the ONUs. In the upstream direction, the splitters also merge signals from the ONUs onto common fibers until the signals from all ONUs merge onto a single fiber in the upstream direction. Signals from the OLT are broadcast to all of its ONUs, and signals from all of the ONUs are received by the common OLT, forming a duplex point-to-multipoint transmission system. The similarity with point-to-multipoint transmission on radio systems means that multiplexing techniques similar to those used for radio can be adapted for use on optical fiber. The passive splitting of the optical fiber creates a fiber tree whose trunk is at the OLT and whose leaves are the ONUs.

The OLT of the passive multistar configuration may operate on a number of fibers, each of which is passively split a number of times before an ONU is reached (i.e., the OLT supports a number of different fiber trees). This may be necessary because the splitting of the optical signal limits the optical power budget, which in turn limits the number of ONUs that can be supported on each tree. Several trees may be needed to provide a suitable level of traffic for a complete optical access network.

The passive nature of the tree configuration of fiber and splitters has led it to being referred to as a PON. The term *optical distribution network* (ODN) is also sometimes used in this context. There has been much debate, mostly well-meaning and some of it well-informed, about the precise definition of the terms PON and ODN. This discussion on terminology has involved issues such as whether the optoelectronic components should be included, whether optically transparent optical amplifiers are really transparent if they are optically pumped, and whether they are active if they are not electrically powered. It is not clear what this discussion of terminology has added to the understanding of the issues.

### 3.3.2 Variations on the Passive Multistar

In addition to the passive multistar configuration, there are some variations that should be noted. Although the passive multistar configuration was described as having a single OLT, there may be occasions when it has two or more OLTs (see Figure 3.9). This would allow the dual-homing of an optical access network (i.e., its connection to multiple host exchanges). This variation increases the reliability of the system in the rare event of the failure of a host exchange or an OLT.

Another possible variation is the passive bus configuration, where a large number of passive single taps are used instead of a small number of many-way splitters (see Figure 3.10). This arrangement has the disadvantage that there must be more points of access to the primary fiber. However, it has the compensating advantage that it can simplify the multiplexing of signals, because the return reflection from the far end of the bus can be used to synchronize the upstream transmission from the remote ONUs. An optical signal injected after the tail of the reflection from the far end would reach the OLT at a specific time independently of the tap at which the signal was injected. *Time division multiple access* (TDMA) can be achieved simply by noting the time at which the reflection from the far end passes the tap and adding a delay before transmissions corresponding to the time slot used by the ONU at that tap.

A further variation on the passive multistar architecture is to use electrically controlled interferometers at the remote ends instead of lasers, since they are potentially less expensive and require less power, reducing the op-

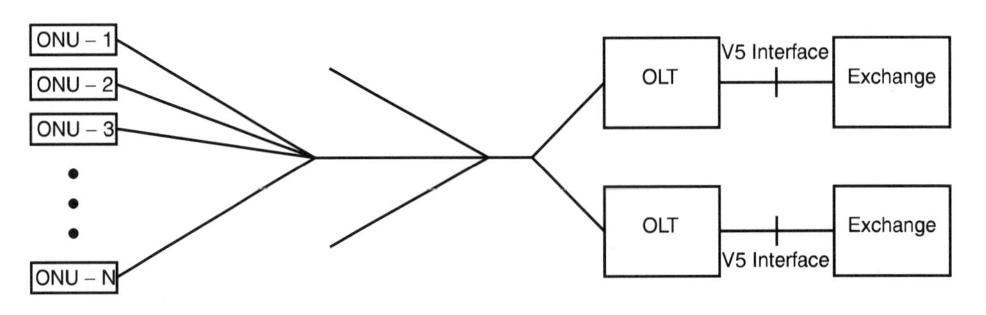


**Figure 3.9** Dual-homing with two OLTs.

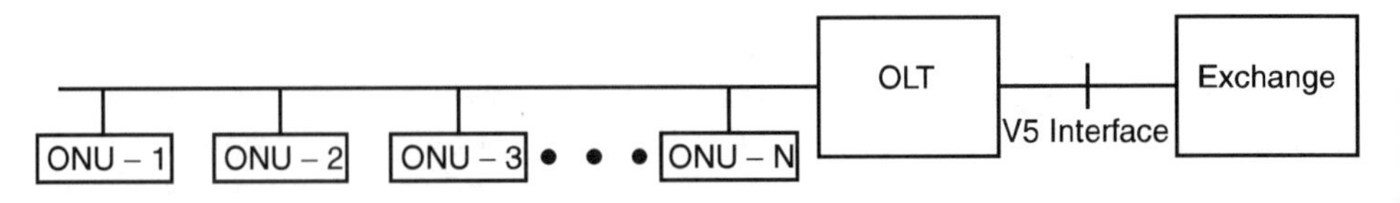


**Figure 3.10** Passive bus fiber topology.

erational problems of powering. This variation results in a reduction in the number of passive splits, because the optical power available for reflection is reduced by distribution downstream through the splitters in addition to the normal loss due to the splitters in the upstream direction. Despite the reduced split, this option is attractive because only two laser diodes are needed for all users, and these can be housed in the hopefully benign environment of the OLT. One laser is used for the modulated downstream transmission to the ONUs and the other is the source that is modulated by reflection at the ONUs and returned to the OLT.

### 3.3.3 Configuration at the ONU

The ultimate goal of optical access technology is to provide a fiber link into every home. This simple configuration is known as the *fiber-to-the-home* (FTTH) configuration. The goal of FTTH suffers from the cost penalty of requiring a separate ONU for every customer, even if the customer only requires a single narrowband telephone. It also has the disadvantage that it limits the number of customers per PON, because there is additional loss of optical power by splitting in comparison with systems with several lines per ONU (see Figure 3.11).

The initial cost overheads are reduced if the cost of the enclosure, power supply, multiplexing components, and optoelectronic components at an ONU is shared over several lines. This makes *fiber-to-the-business* (FTTB), *fiber-to-*

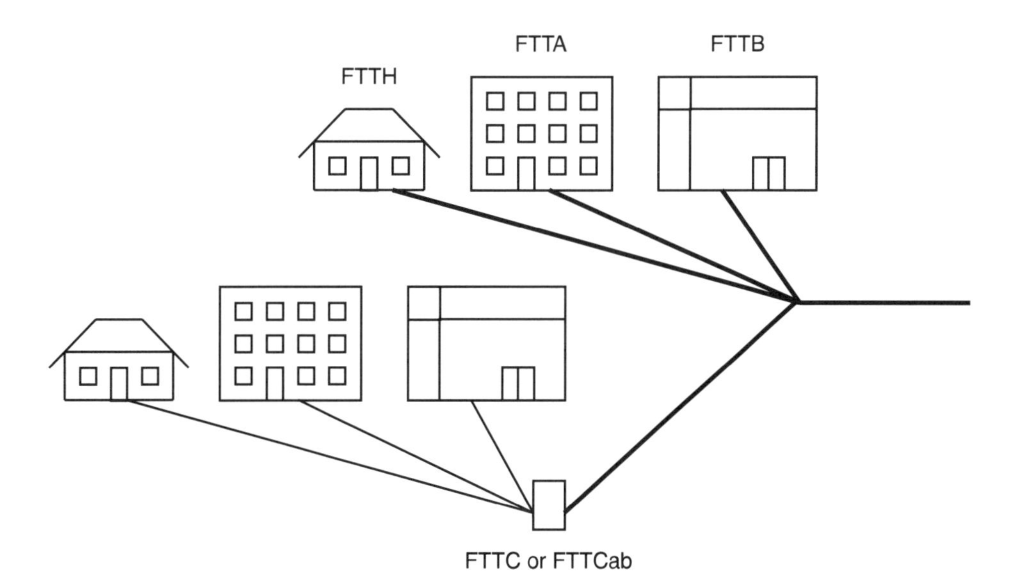


**Figure 3.11** Configurations at the ONU.

*the-apartment* (FTTA), and FTTC configurations more attractive in terms of initial costs than FTTH. These apparently more attractive alternatives are all technically hybrid fiber-copper systems, with slightly more sophisticated multiplexing being performed in the ONUs, but the final copper drops can be kept short and simple. A variation of FTTC is FTTCab, which can reduce the number of ONUs required by making use of more and longer copper drops.

An FTTB arrangement at the remote end takes advantage of the fact that most business customers typically require several lines. This is true for medium to large business customers, although small business customers resemble certain residential customers. An ONU installed at a business location is also likely to be in a secure and benign environment, where the need for backup power may already be recognized, for instance, if a *private automatic bank exchange* (PABX) is already present and a primary supply of mains power is readily available.

FTTA is more appropriate in countries where residential apartments are more common than in those countries where single-family residencies predominate. The environment is similar to that for FTTB, although it may be less secure and backup power may not be as readily available. As with FTTB, there is likely to be an easily available source of mains power, and there may already be a room or enclosure to house equipment within the apartment building.

The FTTC and FTTCab configurations are the most difficult of the variations that share the cost of the ONU over several final network terminations. Unfortunately, they are the ones for which there is potentially the highest demand in countries where most residencies are single-family homes. The pedestal, hole, or cabinet where the ONU is located is subject to temperature extremes. There may not be a mains supply readily available, and even if there is, concerns about flooding and gas accumulation may lead to restrictions about the use of an ONU underground. Perhaps it is fortunate that in the long term these configurations may be superseded by FTTH as increases in volume reduce the cost of ONUs.

### 3.3.4 Hybrid Fiber/Coax

It is important to cable operators that for little additional cost the new networks they are building can evolve to support interactive services for voice, data, and video. Traditional *cable television* (CATV) networks use coaxial cable with long cascades of amplifiers to compensate for signal attenuation. These amplifiers are a reliability hazard, since the probability of all the amplifiers in a long cascade operating properly is significantly less than the probability of a single amplifier operating properly. Even worse, more amplifiers are needed as more bandwidth is used for coaxial transmission, because the attenuation of coaxial cable increases with frequency. In contrast, the attenuation on an optical fiber is independent of the rate at which the carrier light is modulated.

The nonlinear characteristics of laser diodes make them better suited to binary (on/off) transmission over optical fiber than to amplitude-modulated transmission. However, the low attenuation of optical fiber in comparison with coaxial cable meant that there was an advantage in developing amplitude-modulated fiber transmission at radio frequencies, because it allowed the reduction of operating costs, since fewer amplifiers were needed and because it avoided the need to convert between different types of modulation. Following this, it was natural to use special fiber transmission, which is compatible with coaxial transmission, to devise new network architectures that are able to support new interactive cable services and yet remain compatible with existing cable services.

The result was the development of the *hybrid fiber/coax* (HFC) architecture. The final delivery of broadband services to customers' homes over a coaxial bus or small tree and branch architecture is combined with star and ring fiber architectures, which provide low attenuation transmission over longer distances (see Figure 3.12). Perhaps the most important point of this combination is that it supports the future evolution of the network, since it has a simple dedicated fiber link in and out of a small serving area. This avoids the limitation of the traditional cable architecture, where the large tree and branch approach meant that all of the customers on the network would have to contend for the limited bandwidth available.

The optical transmission in an HFC network is quite different from that used for more traditional optical communications. Instead of switching the laser on and off, in HFC networks, a broadband radio frequency signal modu-

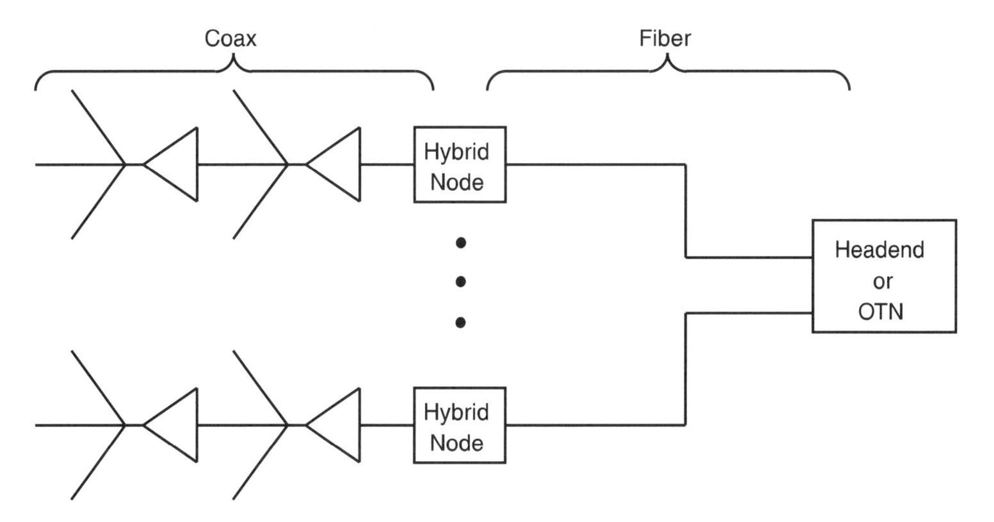


**Figure 3.12** Hybrid fiber/coax architecture.

lates the intensity of a highly linear laser. The broadband signal, which may contain a large number of independent channels, can be sent to a hybrid node of a serving area, where it is easily converted to an electrical signal and distributed to homes over coaxial cabling. The number of homes served by a hybrid node for initial systems is about 500, which is in sharp contrast to the 10,000 to 20,000 homes traditionally served by coaxial cabling. As the demand for bandwidth increases, the number of homes served by the hybrid node can be reduced, possibly to about 100.

In an HFC network, the headend that traditionally served 10,000 to 20,000 homes over a tree and branch cable network may make use of a number of intermediate *optical transition nodes* (OTN) to reach the remote hybrid nodes. The reason is that this reduces the number of fibers needed between the headend and the hybrid nodes. The OTNs may be linked to the headend by a fiber star or ring, and the headends themselves may be connected by a fiber ring to a master headend to form a complete managed broadband network. An SDH ring is an obvious choice to join the headends because it can make use of existing equipment and standards and because the ring provides protection against the failure of one of its links. There is a clear similarity here with the use of a feeder transmission system for a PON-based access network.

One of the strengths of the HFC approach is that it maintains compatibility with existing cable transmission techniques while having the potential for future upgrades. Future cable services will include digital video services, which will normally use compression techniques to reduce the data rate. For cable services, compression is especially attractive because less bandwidth is needed for compressed digital transmission than for traditional analog transmission. However, it should always be remembered that the techniques used for digital transmission by cable operators on HFC networks differ from those used by telecommunications operators on PON networks, because HFC networks make use of linear modulation techniques, while PONs make use of nonlinear binary transmission.

Ideally, there would be a common agreed-on approach for future interactive services on HFC, possibly ATM-based and using linear coding of the ATM data with time-division multiplexing for the point-to-multipoint transmission. Realistically, a number of different proprietary approaches are likely to be used, but one of the advantages of HFC is that these can be accommodated through the allocation of appropriate spectral ranges. The services themselves may be provided at master headends, at remote headends, or even at OTNs as appropriate, and this may change as the network evolves.

Data services are seen as potentially even more attractive to HFC operators than interactive video services. Cable modems that can offer megabit per second data rates to customers in contrast to the tens or hundreds of kilobits offered by conventional modems are the basis for data services. Typically, data modems use TDMA on a specified frequency band of an HFC network. Although the first generation of cable modems has commanded premium prices for proprietary implementations, the demand has been significant, and further generations are foreseen as dropping in price to near the price of telephony modems, since the technology is not intrinsically expensive. Internet routers provide the gateways for traffic to other locations and the placing of Internet routers within an HFC network critically affects the performance of cable modems, since the performance of cable modems decreases as more modems share the same frequency band.

### 3.3.5 General Architectural Issues

So far it has been implied that only a single fiber is required for each ONU in an optical access network. This is true for many types of optical transmission systems, but not for all. Some require two fibers to support duplex operation, with significant implications for both operational and maintenance costs. One of the common maintenance problems with copper is due to split copper pairs, and confusion over which two fibers out of the many in a cable form a single fiber pair may be just as common as confusion over the same issue for copper pairs. Although optical fiber does not suffer from crosstalk as copper does, the use of fiber pairs doubles the amount of fiber that needs to be installed, requires more elaborate record keeping and fiber identification, and increases labor costs. There may be some advantages due to future flexibility with two-fiber operation, but this appears to be more of a rationalization rather than a reason for two-fiber operation.

There is a general consensus on the wavelengths used for transmission on optical access networks. There are two windows of operation on optical fibers, both in the near infrared. One is around 1.3 μm and the other is around 1.5 μm. It has been generally agreed to use the 1.3-μm window for initial systems supporting narrowband services, and to reserve the 1.5-μm window for broadband upgrade. This allocation has the advantage that optical amplifiers are more readily available at 1.5 μm, where they can help maintain the photons per bit required for higher speed broadband operation.

It is often said that the information-carrying capability of optical fiber is almost limitless, and figures on the order of $10^{12}$ bps are sometimes quoted. This suggestion is far from the truth, because it does not take account of the reality of optical transmission. In practice, an optical receiver must receive on average a certain number of photons per transmitted bit. The amount of optical power transmitted must be increased if the data rate is increased, because less time is available for the same number of photons to be received. At sufficiently high levels of optical power, the fiber begins to exhibit significant nonlinear effects, which impose limits on the transmission rate.

The need to maintain an adequate number of photons per bit also constrains the ability to upgrade an optical access network to operate at higher data rates. A PON that has a splitting factor that permits narrowband transmission with a reasonable margin in its optical power budgets cannot support broadband operation if the optical power levels are not boosted in some way, because sufficient photons cannot be received in each bit period (see Figure 3.13). In theory, the number of photons required per bit can be reduced through coding, but in practice, upgrading to a higher bit rate is likely to require the use of optical amplifiers.

The reach of an optical access network can be much greater than that of a copper pair, because the attenuation of optical fiber is much less than that of copper. The OLT of an optical access network can also be sited at a remote location and connected to its host exchange by a feeder transmission system or network (see Figure 3.14), making the remote OLTs satellites of their exchange. A feeder network also simplifies the connection of a single remote OLT to different exchanges, without the need for a second OLT or separate transmission to a different exchange.

Siting the OLTs at remote locations appears similar to using an active multistar, but the degree of the problems is less because the host exchange can now serve many more customers. Additional maintenance costs are not in-

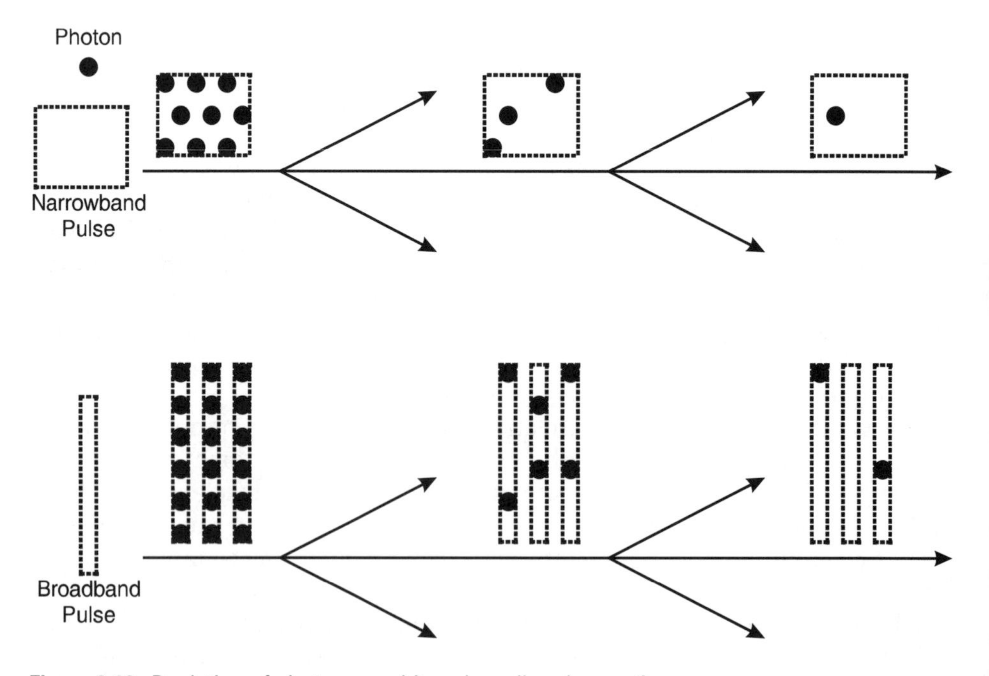


**Figure 3.13** Depletion of photons per bit on broadband operation.

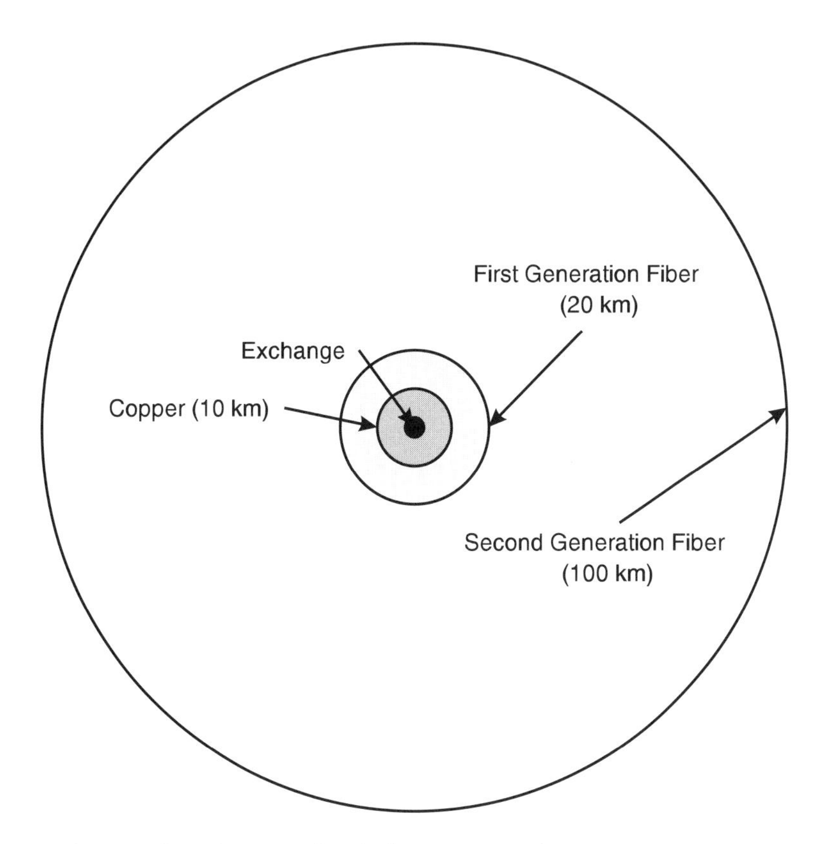


**Figure 3.14** Increased serving area of optical access networks.

curred, and in fact the maintenance costs of the network can be reduced, because the OLTs can be used instead of exchanges. The cost of maintaining the OLTs is less than the cost of maintaining the exchanges because the amount of equipment is less, and the capital cost of an OLT is also less.

The use of remotely located OLTs instead of exchanges amounts to a restructuring of the entire traditional telecommunications network. A new structure could consist of a small number of large exchanges in the core that are served by an outer mantle of feeder networks and optical access networks. A further enhancement, similar to the change from an active multistar to a passive multistar, is possible if optical amplifier technology and wavelength-division multiplexing technology is used to integrate the optical access networks with the feeder networks making use of passive wavelength-dependent splitting. This enhancement would change the exchange ends of the feeder networks back into OLTs and eliminate the separate remote OLTs and the remote ends of the feeder networks.

## 3.4 DUPLEXING AND MULTIPLEXING

A passive multistar architecture requires a duplex point-to-multipoint transmission system, because there is transmission both downstream from the OLT to all of its ONUs and upstream from all of the ONUs to the OLT. The downstream transmission can be relatively simple, since the OLT can broadcast a multiplexed signal to all of the ONUs, which can then select their relevant channels. Achieving bidirectional (duplex) operation on a single fiber is slightly more difficult, because a technique to distinguish between the directions of transmission is required. Coordinating the multiplexing of the upstream transmission from all of the ONUs to the OLT is a little harder again.

The terms *duplexing* and *multiplexing* are sometimes not clearly distinguished in the terminology used in this area. In particular, the term *multiplexing*, which indicates how a number of channels are carried together, is sometimes used to indicate how channels are carried in different directions, which is more accurately described as *duplexing*. The reason for this ambiguity in terminology is that certain multiplexing techniques that can be used to differentiate between channels traveling in the same direction can also be adapted to differentiate between channels traveling in opposite directions. Although an attempt will be made to use the correct terminology, the use of *multiplexing* to mean *duplexing* is so widespread, especially for electronic duplexing techniques, that the use of correct terminology could be confusing, especially when consulting other published material.

Optical heterodyne and homodyne techniques will not be considered here, because the components are not sufficiently robust or easily available to be used for initial forms of optical access networks, and it is also possible that advances in optical amplifier technology will make these techniques unnecessary. This is discussed in more detail in the next chapter.

### 3.4.1 Duplexing Techniques

A number of techniques may be used to separate the directions of transmission on an optical fiber. These range from the use of a different fiber for each direction, through optical techniques using directional coupling and different wavelengths, to the use of electronic subcarriers and time multiplexing.

#### *3.4.1.1 Fibers*

The simplest way of separating the directions of transmission is to use a different fiber for each direction. This technique is referred to as *space-division multiplexing* (SDM) or, more accurately, *space-division duplexing* (SDD). Although this approach is very simple, it has several disadvantages. It has higher cable costs because twice the amount of fiber is required. The installation costs can also be higher because additional cables will need to be installed more often and because more time must be spent connecting the fibers and finding two unused fibers instead of one. More complex records also need to be kept to identify both of the fibers for each link, and this further adds to the costs.

SDD is an inefficient way to use the capacity of the fiber, but at least it works. The reason for its use in some systems is the failure of the approach that was originally intended. This alternative approach is *directional-division multiplexing* (DDM), which is more accurately called *directional-division duplexing* (DDD).

#### 3.4.1.2 Directional Coupling

The DDD approach is based on the use of directional couplers that make use of technology similar to passive splitters. The bidirectional signals on an optical fiber can be separated into unidirectional signals on two fibers to which it is coupled provided that signal is lost into a fourth leg of the coupler. The normal reversibility of the direction of travel of light is not violated, because reversibility would require the lost light to be returned from the fourth leg of the coupler. The received optical power is reduced by about a factor of 4 if directional couplers are used, because about half of the power is lost at each end. The corresponding electrical power after optoelectronic conversion is reduced by a factor of 16.

The DDD approach failed because of reflections. The various splitters and couplers all produce optical reflections due to mismatches between components. This should not have been surprising, because copper pairs also give rise to reflections in the same way. Unfortunately, the reflections of the transmitted signal interfere with the signal received from the far end of the PON, which is attenuated by the passive splitting. It was originally hoped that the reflections could be minimized in the fabrication of components, but this proved to be a classical example of wishful thinking. Transmission systems designed to operate using DDD are most easily adapted to use SDD by removing the directional couplers and using two fibers.

DDD is not likely to re-emerge in the future, because the infrastructure of the fiber network would need to be upgraded to use higher quality splitters. It could be used in new systems if lower loss components are available. Even with existing components, echo cancellation could be used to remove the degradation due to optical reflections, although it would need to operate at rates that are higher than on copper systems. But despite these possibilities, DDD would still have a disadvantage because of the loss of optical power and because alternative approaches are readily available.

#### 3.4.1.3 Optical Wavelengths

Using different optical wavelengths is possibly the most obvious way to separate the directions of transmission on an optical fiber. The fiber can carry a wide band of wavelengths, which can be separated by optical filters. This has been referred to as *wavelength-division multiplexing* (WDM), *wavelength-division duplexing* (WDD), *optical frequency-division multiplexing* (OFDM), and *optical frequency-division duplexing* (OFDD). There are two windows of operation on optical fiber, one around 1.3 μm and the other around 1.5 μm. The simplest approach is to one window for each direction. Unfortunately, this simple approach is in conflict with the view that the 1.5-μm window should be reserved for upgrading to broadband operation.

A more sophisticated approach is to use *high-density wavelength-division multiplexing* (HDWDM) to split up a single window into several bands. This can be more expensive because it requires more elaborate filters and more precisely fabricated lasers. These components are not likely to be appropriate for initial transmission systems for narrowband services, since alternatives are more readily available. The components are likely to be used in later systems, especially for broadband operation, but they are more appropriate for multiplexing rather than for duplexing, because optical bands are a limited resource and other techniques for duplexing can be used.

#### 3.4.1.4 Pseudorandom Electronic Modulation

The remaining duplexing techniques are electronic in nature. These use TCM, *subcarrier multiplexing* (SCM), and *code-division multiplexing* (CDM). The CDM approach is not applicable, because in general it requires pseudorandom modulation to be performed at a rate much higher than the operating rate, and the hailed advantage of gentle degradation as the number of channels increases is not relevant to duplex operation, since only two channels are to be separated. If CDM is simplified and made specific for two-channel duplex operation, then it becomes equivalent to SCM.

#### 3.4.1.5 Electronic Spectra

The SCM approach is sometimes called *frequency-division multiplexing* (FDM) because it can be thought of as frequency-division multiplexing over an optical ether. It is better to use the term SCM, because this indicates that there is an electronic subcarrier in addition to the optical carrier, whereas the term FDM can lead to confusion between optical and electronic frequency domains.

For duplex use SCM can be simplified to baseband transmission in one direction and transmission in the opposite direction, which is shifted in electrical spectrum only enough for electronic filtering to separate the two signals.

High-speed components are especially not required because biphase modulation is sufficient to shift the electronic spectrum far enough to allow directions to be separated.

The spectrally shifted transmission can be chosen for either upstream or downstream use. There may be a marginal cost advantage of a specific choice because more components are used in the ONUs where either higher speed receivers or higher speed transmitters are required, depending on the choice. If electronically controlled reflectors are used instead of lasers in the ONUs, then there may be an advantage in shifting the downstream transmission out of the baseband region, because this decreases the background noise with which the modulated reflection from the remote ONUs must be compared. The higher speed modulation spreads the optical spectrum transmitted by the laser in the OLT and reduces the impact of coherent addition of Raleigh reflected light back to the OLT.

#### *3.4.1.6 Time Intervals*

The TCM approach is an optical version of the ping-pong technique, which is also used on certain copper DSL transmission systems. This approach is better suited to use on optical fiber than on copper, because fiber does not attenuate more at the higher rates of operation required in the way that copper does. These rates of operation are no greater than those required for the SCM approach and are less if SCM uses *frequency modulation* (FM) rather than *phase modulation* (PM) since frequency modulation requires higher speed optoelectronics.

The TCM approach has to take account of the optical transmission delay in the fiber, because the OLT normally has to wait after each burst until it receives a burst from the remote ends before it sends its next burst. The overhead of lost time due to the transmission delays can be reduced if the bursts are longer, because this increases the ratio of active transmission to transmission delay. Unfortunately, lengthening the bursts increases the overall transmission delay, because the transmission must be buffered during the longer interval while the opposite end is transmitting. The transmission delay cannot be increased indefinitely, since this would affect the quality of service.

The range of operation of TCM is limited by optical transmission delay in this way unless a more sophisticated *multiball ping-pong* approach is used. In practice, this is not a significant problem, because the range of *single-ball ping-pong* is adequate for the initial deployment of systems. TCM has the advantage that it is unaffected by near-end optical and electronic crosstalk, since the receiver is inactive during the periods when it could detect a transmitted signal. It also has a second and major advantage that it can be implemented simply with only digital electronics.

### 3.4.2 Multiplexing Techniques

Techniques used to separate the directions of transmission can often be adapted to separate different channels transmitted in the same direction. The exceptions to this generalization are techniques that use separate fibers (SDM) or different directions (DDM). It is absurd to consider using SDM, because this would mean that different fibers were used for different channels, which is not really multiplexing. DDM cannot be used either because it uses the direction of transmission to separate channels, while the channels to be separated are transmitted in the same direction. SDM and DDM are duplexing techniques and not true multiplexing techniques.

Duplexing techniques that separate the channels using optical wavelengths, pseudorandom modulation of an electronic carrier (CDM), electronic spectra, and time intervals can be used for both duplexing and multiplexing. CDM will not be considered further, because it is even less well suited to multiplexing than it is to duplexing, since very-high-speed components are needed, as it cannot be simplified in the same way as for two-channel operation.

#### 3.4.2.1 Optical Wavelengths

If different optical wavelengths are used for different channels, then HDWDM must be used, because otherwise only two channels could be multiplexed, since there are only two optical windows. The same considerations of cost and availability apply to the use of HDWDM for multiplexing as for duplexing.

HDWDM is likely to be used in the longer term to allow different services to share the same fiber, because the high bandwidth available at each wavelength makes HDWDM more suitable for differentiating between different services than between different channels of a single service. It is likely to be used in the 1.5-μm window, since the 1.3-μm window is likely to be used indiscriminately for first-generation narrowband systems. HDWDM allows more effective use to be made of the transmission capability of the optical fiber, but the cost and availability of components make HDWDM less suitable for the first generation of optical access networks.

#### 3.4.2.2 Time Intervals

For downstream transmission, *time-division multiplexing* (TDM) is the simplest solution. It is easy to implement with digital electronics and the continuously broadcast transmission of TDM allows the ONUs to synchronize with the OLT. TDM is more flexible than other approaches, because it allows the basic channels to be small and yet permits them to be easily concatenated into higher bandwidth channels. SCM and WDM approaches are both less well suited to the transport of a large number of small channels and less adaptable for the concatenation of channels.

The difficulties with TDM are mostly on upstream transmission because of the need to synchronize the different ONUs so that they can transmit to the same OLT. The same problems for TDMA occur on optical access networks as for radio systems, and can be solved in a similar way. The key to successful TDMA operation is the ranging protocol, which allows the ONUs to take account of the different transmission delays to the OLT and so ensure that transmissions do not overlap (see Figure 3.15).

Ranging is typically a two-stage process. Coarse ranging must be performed when an ONU is powered up because it may have no knowledge of its associated transmission delays. Ranging also needs to be continuously fine-tuned during operation to take account of temperature fluctuations. The ranging process needs to be fail-safe, because a single ONU transmitting without proper synchronization could disrupt the operation of all other ONUs.

Ranging synchronizes the ONUs so that the OLT receives a continuous stream of synchronized bits. The OLT can then use a single continuous clock to sample the data from the ONUs. In addition, the optical transmitters at the ONUs are less likely to overheat, because bits from different ONUs can be interleaved, giving time for transmitters to cool down between the transmission of successive bits. Unfortunately, the level of required resolution is more difficult to achieve as data rates increase and bit periods decrease.

When the ranging process is unable to synchronize the different ONUs to within a small fraction of a bit period, then ONUs must transmit in bursts with added synchronization bits at the beginning of bursts and additional guard times between bursts. The OLT uses the synchronization bits to decide the best

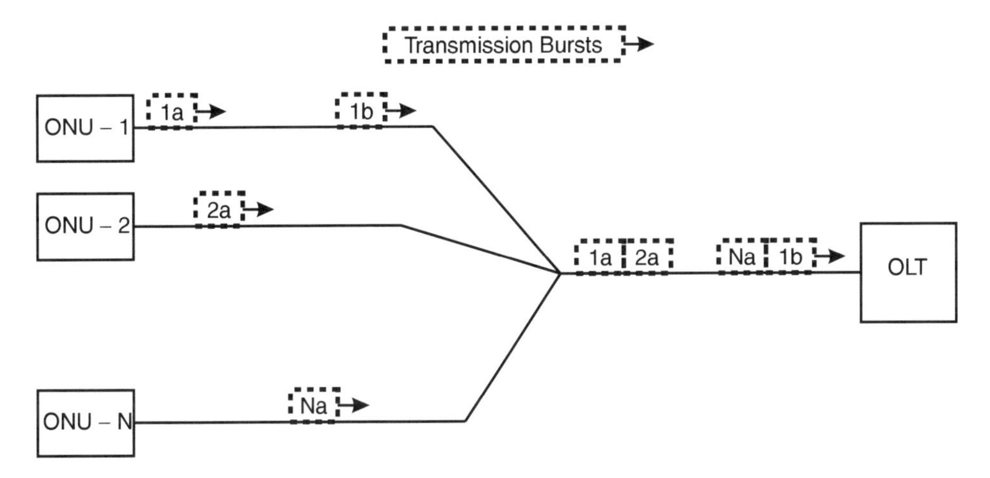


**Figure 3.15** Merging of ranged transmissions from remote ONUs.

place to sample the subsequent bits of the burst, adapting the sampling clock toeach burst. The period of the bursts must be kept low, since it introduces delays caused by the data for transmission being buffered at each ONU between its successive bursts.

Ranging protocols are easily capable of synchronizing ONUs, which have distances to the OLT that differ by up to 20 km. Even more flexible ranging protocols are feasible, since the value of 20 km, which has been implemented in real systems, is not due to a fundamental limitation. The flexibility of the ranging protocol gives optical access networks a greater catchment area than is possible for traditional copper pairs, which are limited by attenuation and loss of line power to about 10 km.

The frames and multiframes typically contain management and control channels in addition to the payload transmission. In addition to ranging, these channels are used to allocate time slots, to convey alarms, and to control testing. Certain of these functions are performed by the system as a whole and others are performed per ONU.

Downstream TDM operation and upstream TDMA operation can be combined with other types of multiplexing, typically SCM. Different types of multiplexing can be used in the different directions of transmission because the different directions of transmission pose different technical problems. Alternatively, these different techniques can be combined in a single direction to create a hierarchy of multiplexing that makes use of the lower granularity and increased flexibility of TDM and TDMA so that services with a wide range of bandwidth requirements can be accommodated.

#### *3.4.2.3 Electronic Spectra*

Different channels can be multiplexed into different electronic spectral bands for transmission in one direction. The allocation of the subcarrier frequencies can be more sophisticated for multiplexing than for duplexing, because subcarrier frequencies may be allocated on demand, as for radio systems. This has the advantage that it can reduce the electronic bandwidth required for the optoelectronic components. It can also reduce the optical power levels, because less power is necessary for operation at the lower data rates. Allocation of frequencies on demand is more complex when SCM is used for duplexing, because both directions are normally used simultaneously.

This use of an electronic subcarrier for multiplexing is easily combined with its use for duplexing. In this application, the baseband can be used in one direction with the TDM of different channels. The other direction can be identified because it uses subcarriers instead of baseband and different subcarriers are used for different channels. However, there can be unintended coherent beating between received signals and reflections of transmitted signals if both directions share the same optical bandwidth and fiber.

SCM is intrinsically more complex than TDM because it is not a purely digital approach. This is not a great disadvantage, since the SCM approach can take advantage of radio technology. SCM has an advantage over TDM for upstream transmission in high-speed optical access networks because it avoids the synchronization problems of high-speed TDM.

## 3.5 UPGRADING TO BROADBAND

Although the initial use of optical access networks is for PSTN, narrowband ISDN, and leased-line services, the ultimate goal is to provide broadband services. In the long term, an optical-fiber or coaxial infrastructure is necessary for broadband services, since copper pairs cannot deliver the bandwidth ultimately required over the necessary distances and radio bandwidth is a limited and valuable resource. It is sensible to be able to upgrade the optical access networks initially installed to support narrowband services, because this avoids the costs of installing a second fiber infrastructure for broadband services.

The first generation of optical access networks do not emphasize FTTH because the equipment costs are seen as being too great. Residential broadband services can be included later either by adding FTTH or by upgrading the final drop of the other residential configurations for the ONU. The direct FTTH approach may be preferable, because it may cost less for broadband services than upgrading the final drop of the other configurations, especially when maintenance costs are taken into account, and it does not create an electronic bottleneck for services. An FTTCab approach has a slight advantage over FTTC because it reduces the number of ONUs where maintenance has to be carried out, but both suffer from the presence of the electronic bottleneck. Business customers will typically require different broadband services to residential customers, and these services can be supported on the fiber infrastructure for the FTTB configuration, since the fiber for this terminates in the business customers' premises.

With an infrastructure in place, it is necessary to have some way of differentiating between broadband and narrowband transmission on the same fiber. Although the same techniques that are used for multiplexing can be applied to differentiate between services, in practice the choice is limited because an infrastructure that is well suited to narrowband services has problems delivering sufficient photons per bit for broadband services. This difficulty can be solved by using different optical wavelengths for the different services with optical amplifiers to boost the broadband signals.

Although the same basic approaches to duplexing and multiplexing apply to both narrowband and broadband transmission systems, the details of the transmission systems differ. The problems of ranging in TDMA operation are greater for broadband transmission. This makes the cells of ATM technology a natural approach, because transmitting them in bursts simplifies the ranging. Alternatively, SCM can be used in the upstream direction instead of TDMA, because this eliminates the need for synchronization and hence ranging for the ONUs.

There was much eager experimentation over the modulation techniques for video-based broadband services. Digital transmission unsurprisingly emerged with the honors. Analog modulation was proposed because it is used for video by certain cable operators and because it is compatible with the vast number of TV sets, despite the fact that the nonlinearities of optical components make analog operation more difficult. Frequency modulation was also proposed and is more justifiable because it is less dependent on linear operation and allows existing set-top boxes to be used.

Digital video transmission has won over frequency modulation because the quality of transmission can be guaranteed, since the error rate can be monitored and digital electronics can be more easily shrunk and cost reduced. It is also easier to scramble a digital signal to restrict the access. Standards for video compression have also helped significantly by reducing the bandwidth required by more than an order of magnitude.

The possibility of interference between broadband and narrowband systems operating on the same fiber should not be discounted during an upgrade. Both optical and electronic crosstalk can occur between different systems, and even if they are minimized, it should be remembered that optical fiber has nonlinear properties and does not have practically infinite bandwidth.

It should also be noted that the assumption that optical access networks will be initially installed for narrowband services can be questioned. It may be more effective to use the existing copper infrastructure for narrowband services and to experiment with advanced copper transmission for broadband services, because it is less of a risk than installing a narrowband optical access network. In this case, a broadband optical access network could be installed from scratch when there is a sufficient market for broadband services to justify it, without the problems of needing to support emergency narrowband service.

## 3.6 COMMON FALLACIES

The operational aspects of optical access networks have been a topic rich in confusion and fallacy, a tribute to human fallibility and the triumph of creativity over veracity. The wisest advice may be to approach the entire area, including the discussion here, with a level of skepticism that borders on paranoia.

In fact, paranoia itself is part of the story. Serious concerns were voiced about the security of transmission on optical access networks, because the downstream transmission to all ONUs can be detected by tapping the PON. This is not significant, because it is far easier to tap a copper pair than to tap an optical fiber, but this argument only fed the paranoia because it emphasized the lack of security on traditional copper pairs.

A proposed solution to this perceived threat was to ensure that only properly authorized and authenticated ONUs should be connected to a PON, and a challenge and response mechanism has been proposed similar to the friend-or-foe identification techniques used for military aircraft. This suggestion contains two problems. First of all, the lack of security on copper pairs is not a valid reason to improve the security of fiber transmission systems, because the issue in question is the intrinsic security of the fiber transmission. Secondly, even if it were necessary to improve the security of fiber transmission, a challenge and response mechanism is not the correct approach, especially since tampering can already be detected by its effect on ranging and leveling.

The techniques used for military aircraft are based on the assumption that a military aircraft is unlikely to be readily available for an enemy to analyze. Unfortunately, ONUs are very likely to be available for analysis, because an ONU is somewhat easier to steal than an advanced military aircraft.

Another operational fallacy involves the much heralded blown-fiber installation technique. The idea here was to install plastic tubing in ducts as a matter of course and then to install fiber later by blowing it in later along the tubing. The perceived advantage was to prepare for the installation of a fiber infrastructure while saving on the cost of the fiber. Unfortunately, this incurs two sets of installation costs, once for the tubing and once for the fiber itself. It can be more effective to simply install the fiber at the onset because this avoids both the operational costs and delays of blowing the fiber and the cost of the tubing. There may still be some advantages to blown fiber, but it is unlikely to replace conventional cabling techniques.

The powering of the remote ONUs is also major operational issue. This issue is so significant that it is discussed for access networks in general in a separate chapter. Unfortunately, the widespread recognition that a problem existed seems to have lead to the view that someone must have solved it. The possibility that there may be a serious difficulty that might interfere with the widespread deployment of optical access networks cannot be entertained by anyone with a properly positive attitude and a dismissive attitude towards inconvenient technical details.

Fortunately, there are operational benefits to optical access networks. There have to be operational benefits because the equipment costs cannot be lower than for copper, since more equipment is needed. These operational benefits are due to more effective operational procedures that can be realized with optical access networks. Unfortunately, there is a bit of a problem with the management interface necessary to support these procedures.

The accepted wisdom on management interfaces is not compatible with implementational simplicity. Although the cost of a management interface is high, this is not the greatest problem, since it is possible for a single interface to support a large number of customer ports. The more significant factor is the level of the investment required to develop the management interface, since this can be greater than the development cost of the underlying functional system. A supplier who makes this level of investment without safeguards is gambling with the future of the company.

## 3.7 SUMMARY

The story of optical access technology is the story of the search for a cost-effective way to deploy optical fiber in the access network and for services to justify it. The fundamental structure of a telecommunications network could change if optical access networks could be economically deployed in conjunction with feeder networks or in large passive multiwavelength architectures.

Monomode optical fiber with the light confined by total internal reflection is preferred, since it avoids the spreading of optical pulses that occurs with multimode fiber. Splitters and directional couplers can be constructed by making use of the tunneling of photons between adjacent optical waveguides. Filters, which are necessary to prevent photodiode receivers from responding to undesired wavelengths, may be constructed by creating a number of thin optical layers with different refractive indexes.

Laser diodes are preferred to LEDs as optical transmitters, because their coupled optical power is greater and their narrower bandwidths prevent pulses from broadening due to chromatic dispersion and allow the fiber to support operation at different wavelengths. Photodiode receivers use a reverse-biased p-n junction. For very sensitive operation, a PIN photodiode with an optical preamplifier has advantages over avalanche photodiodes and coherent detection, and the advent of simple and robust optical amplifiers based on fiber doped with rare earths will revolutionize optical access networks in the longer term. There are advantages to using dual-mode transmitter/receiver diodes if an optical ping-pong approach is used to separate directions of transmission. Reflective modulators can avoid the need for laser diodes in the ONUs but limit the multistar fan-out. On all optoelectronic components, cost is a critical factor and high-volume markets are required for costs to be reduced.

Economic considerations have lead to the development of passive multistar configurations with bidirectional point-to-multipoint transmission between a single OLT at the exchange end and a large number of ONUs at the remote ends. At the ONU, the fiber can be terminated at the customer's premises in the FTTB, FTTA, and FTTH configurations. Alternatively, the ONU may be located at either a small or a large distance from the customer's premises, as in FTTC and FTTCab, respectively. If the ONU is not located at the customer's premises then the problems of powering and of the environment at the ONU are aggravated.

The advantages of fiber transmission have also led to the development of HFC networks, where coaxial cable is used for a serving area of 500 homes or less, in contrast to the 10,000 or more homes traditionally served by a coaxial network. These smaller serving areas are grouped together through dedicated fiber links with compatibility between the optical transmission on the fiber and the traditional electrical transmission on the coaxial cable achieved through the use of highly linear intensity-modulated lasers. HFC reduces the contention for bandwidth between customers, which limits the addition of new interactive services to cable networks, and the high bandwidths they support allow different spectral ranges to be allocated to different services.

For the various architectures, separating the directions of transmission within an access network by using different fibers is feasible, but it requires more complex operational practices to keep track of the associated fibers. Using directional couplers to separate directions is not feasible because of the optical reflections from mismatched components, and it is also wasteful of optical power. Directions may also be separated by using a different wavelength in each direction, but this is an inefficient use of the optical bandwidth. The most effective way of separating directions is to use electronic techniques, provided coherent beating can be avoided. These may involve either electronic modulation of the optical signal to shift transmission on one of the directions out of the electronic baseband or optical ping-pong transmission, which allows transmission in the different directions at different times.

The different channels transmitted in the same direction can also be separated by using different optical wavelengths for each channel, but again this is wasteful because the different optical wavebands are a limited resource, and it would be better to reserve different wavebands for different services. For downstream transmission, it is simplest and easiest to use TDM to separate the channel, but TDM is more difficult for upstream transmission from the ONUs, since they need to be synchronized so that transmission from all of them merge smoothly at the OLT despite the different transmission delays, making cell-based upstream transmission or different electronic modulation more attractive as the transmission rate increases.

Only optical-fiber or coaxial cable readily provides sufficient bandwidth for broadband services in the long term, and it should be possible to upgrade narrowband optical access networks to avoid the need for a second fiber infrastructure for broadband operation. This assumes the widespread deployment of narrowband optical access networks, since copper pairs could continue to be used for narrowband services and for initial broadband offerings. For FTTB and FTTH, upgrading is less of a problem, because there is no final copper pair drop to the end customer. The initial equipment costs of FTTH are higher than for other residential configurations, but the operating costs may be less and it has no electronic bottleneck to be overcome when upgrading. The 1.5-μm transmission window and ATM-based transmission will be widely used for broadband services, with 1.3 μm often limited to narrowband unless a separate broadband overlay is used. For broadband upgrade of a narrowband network, optical amplifiers may need to be used.

## Selected Bibliography

Altwegg, L., A. Azizi, P. Vogel, Y. Wang, and F. Wyler, "LOCNET: A Fiber in the Loop System With no Light Source at the Subscriber End," *J. Lightwave Technology*, Vol 12, No. 3, March 1994, pp. 535–540.

Carrol, C., and W. Clement, "Using Hybrid Fibre-Coax Networks for Demand and Inter-active Services," *International Broadcasting Convention 95*, September 1995, pp. 316–331.

Clarke, D. E. A., and C. E. Hoppitt, "The Design of a TDMA System for Passive Optical Networks," *ICC'90 Conf. Record*, Vol. 2, April 1990, pp. 654–658.

Faulkner, D. W., D. B. Payne, J. R. Stern, and J. W. Ballance, "Optical Networks for Local Loop Applications," *J. Lightwave Technology*, Vol. 7, No. 11, November 1989, pp. 1741–1751.

Gobl, G., C. Lundquist, B. Hillerich, and M. Perry, "Fibre to the Residential Customer," *GLOBECOM'92 Conf. Record*, Vol. 1, December 1992, pp. 165–169.

Kashima, N., "Upgrade of Passive Optical Subscriber Network," *J. Lightwave Technology*, Vol. 9, No. 1, January 1991, pp. 113–120.

Mochida, Y., "Technologies for Local-Access Fibering," *IEEE Communications Magazine*, Vol. 32, No. 2, February 1994, pp. 64–73.

Oakley, K. A., C. G. Taylor, and J. R. Stern, "Passive Fibre Local Loop for Telephony With Broadband Upgrade," *ISSLS 88 Proc.*, September 1988, pp. 179–183.

Paff, A., "Hybrid Fiber/Coax in the Public Telecommunications Infrastructure," *IEEE Communications Magazine*, Vol. 33, No. 4, April 1995, pp. 40–45.

Pugh, W., and G. Boyer, "Broadband Access: Comparing Alternatives," *IEEE Communications Magazine*, Vol. 33, No. 8, August 1995, pp. 43–46.

Uneoya, T., F. Ashiya, N. Tomita, K. Satoh, and T. Sakai, "Operation, Administration and Maintenance Systems of the Optical Fiber Loop," *GLOBECOM'90 Conf. Record*, Vol. 3, December 1990, pp. 1493–1497.

# Advanced Optical Technology

# 4

"I don't think we're in Kansas anymore."
—*The Wizard of Oz*

There are a number of recent developments and advanced concepts that may have a significant effect on the future of optical access networks, and even on the structure of the entire telecommunications network. Optical amplifiers are likely to have a key role in the long term, and so may coherent optical systems. The concepts of optical ether and distributed switching could revolutionize the structure of the entire telecommunications network. Overcoming the limiting factors of optical-fiber transmission leads to the possibility of soliton transmission within access networks, and the ultimate form of secure data transmission is offered by the technique of quantum cryptography.

## 4.1 OPTICAL AMPLIFIERS

An *erbium-doped fiber amplifier* (EDFA) is a short length of optical fiber containing a small proportion of erbium ions with a separate pump laser, which excites these ions so that their decay can be used to amplify a suitable optical signal. An optical amplifier differs from a traditional optoelectronic amplifier in that it does not require the conversion of the optical signal into an electronic signal and back again. The pump laser can travel in the same direction as the signal being amplified (copropagating) or in the opposite direction (counterpropagating), although noise and gain performance can be improved if the doped fiber is pumped in both directions at the same time. Typically, optical isolators are needed at either end of the amplifier section to prevent the amplifier from oscillating due to feedback from spurious reflections (see Figure 4.1).

Optical amplifiers have a large bandwidth, which makes them suitable for use in WDM systems, and the lifetime of the excited state of the erbium ions is

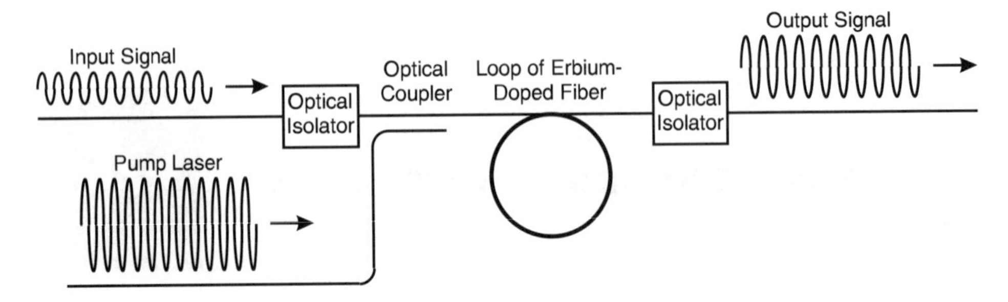


**Figure 4.1** Structure of a copropagating optical amplifier.

long, which gives low crosstalk between WDM channels. Output power of hundreds of milliwatts can be achieved, and gains of up to 50 dB have been obtained with two-stage amplifiers. Each time the optical signal is amplified, noise due to *amplified spontaneous emission* (ASE) is added to the signal. If a chain of amplifiers is used, then the ASE accumulates and is amplified with the signal as it travels down the chain.

An optical amplifier can be used as a power amplifier to increase the optical power when a signal is initially launched down a fiber, or as a repeater to boost a signal that has been diminished through attenuation or splitting, or as a preamplifier immediately before an optical detector.

### 4.1.1 Preamplification

Optical amplifiers are very effective as preamplifiers for receivers (see Figure 4.2), since little loss is introduced through coupling them onto a fiber and very high levels of population inversion can be achieved for the erbium ions. Performance close to the limit predicted by quantum theory can be achieved. Amplification is also independent of modulation and bit rates, and the equip-

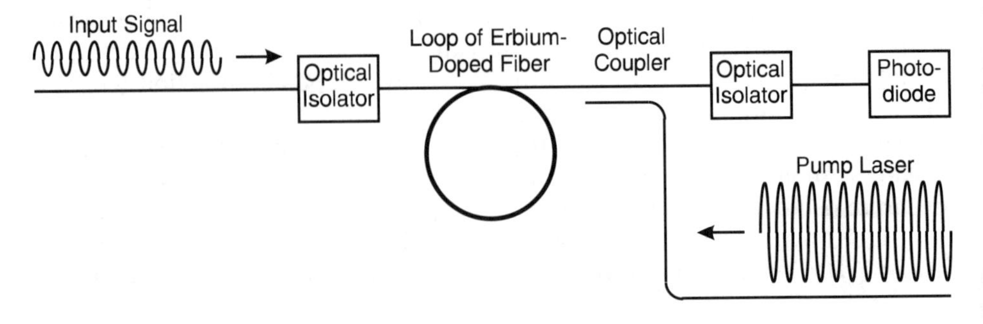


**Figure 4.2** Counterpropagating optical preamplifier.

ment reliability is greater than for optoelectronic amplifiers because the construction is simpler.

A narrowband optical filter is required between the optical preamplifier and the detector to filter out the broad-spectrum noise produced by ASE. Over 20 dB of improvement can be achieved in this way, and the results with a simple low-noise PIN receiver are comparable with those of a coherent optical receiver. The next step in the development of this technology appears to be the integration of the optical amplifier with the PIN diode by creating amplifying waveguides on the same silicon wafer as the diodes instead of using a separate fiber amplifier.

An additional advantage of erbium preamplifiers for optical access networks is that they eliminate the need for additional power units in the field, since the optical pumping of the amplifier can be powered from the same electrical source as the receiver itself. There are advantages to distributing the amplification along the fiber, since it reduces the accumulated noise, but it needs to be offset against the problems of powering in an access network. In any event, the use of erbium preamplifiers in optical access networks is likely to occur after the technology is mature and has been deployed in long-haul routes, particularly for undersea cables.

### 4.1.2 Optical Amplifiers and Regenerative Repeaters

Optoelectronic regenerative repeaters have been traditionally used to compensate for the attenuation of optical fiber over long distances. It is preferable to use nonregenerating optical amplifiers instead, because they are more sensitive and robust. In addition, a single optical amplifier can handle a number of WDM channels with only an increase in the pumping power being needed as further channels are added. Regenerative amplifiers are only necessary when transmission is limited by dispersion, nonlinear fiber effects, or accumulated noise from the optical amplifiers, and it is sensible to recover the digital information and regenerate the original optical signal before further transmission.

The spacing of optical amplifiers in simple applications is principally determined by the attenuation of the optical fiber, since signal levels need to be maintained. For access networks, the attenuation can be dominated by passive splitters, and this may require amplifiers to be located before the points of splitting. Although it is possible to increase the amplifier power to compensate for large attenuations due to long fiber lengths between amplifiers or to large split ratios, this is not effective because it creates disproportionally high amplifier noise. If optical amplifiers are used instead of optoelectronic repeaters, then the spacings between amplifiers required may be reduced slightly because of the noise introduced by the amplifiers, which is not present when the optical signal is regenerated.

## 4.2 COHERENT OPTICAL SYSTEMS

In WDM systems, optical filters are used to separate the channels with channel spacings on the order of 1 nm, giving about 30 channels in the wavelength window around 1.55 µm. These channel spacings correspond to a channel bandwidth of about 100 GHz, and the whole of this bandwidth is unlikely to be fully used for transmission, because this would require extremely high data rates. Coherent techniques can make better use of the 1.55-µm window, because they support a larger number of channels with smaller channel spacing and so are better suited to the achievable data rates. Multiplexing with channel spacings on the order of 1 to 10 GHz based on coherent optical technology is referred to as OFDM to distinguish it from WDM using optical filters.

### 4.2.1 Coherent Receivers

Coherent detection allows narrower channel spacings than those possible with optical filters, because the filtering is done electronically after converting the optical signal to an intermediate radio frequency electronic signal (see Figure 4.3). Coherent detection is also more sensitive than direct detection because it gives a greater variation in the intensity of the detected light, but this advantage is reduced if direct detection is performed after preamplification by an optical amplifier.

Coherent receivers apply the same principle of heterodyne detection that is used in radio receivers to coherent laser signals. Light from a local laser source that is close in frequency to the channel to be selected is injected together with the incoming signal into the receiver diode. This diode outputs an electronic signal that is proportional to the intensity of the incident light. Channels other than that which is to be selected are filtered out by a low-pass electronic filter.

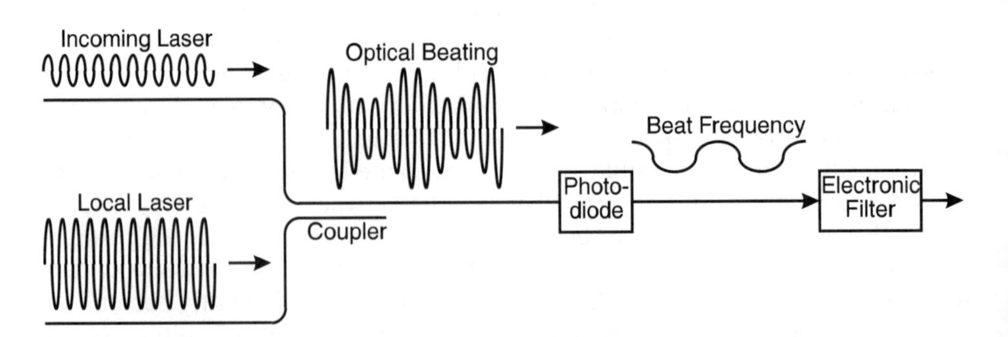


**Figure 4.3** Operation of a coherent optical receiver.

If the coherent light from the local laser is in phase with the received coherent light, then the two are added together, whereas they cancel each other out if they are out of phase, producing a modulation of the amplitude of the detected light. This interference between the two light sources amplifies the variation in the intensity of the light, since the intensity is proportional to the square of the amplitude. This coherent heterodyne detection takes advantage of the coherent nature of the received signal, since the amplitude variations due to noncoherent received light average out, because their phases are random.

Optical heterodyne detection has the advantage over radio heterodyne detection in that no separate mixer is needed, since the detector diode that must be present in any case performs the equivalent function. However, this advantage is offset by the need for a greater level of stability in the lasers to maintain the phase relationships between the transmitters and the local sources and to maintain the spacings between the optical channels. The use of an intermediate radio frequency also makes the receivers more complex.

### 4.2.2 Coherent Transmitters

Transmitters for coherent systems could be modulated by switching the transmitted signal on and off and using the presence or absence of the intermediate radio frequency in the receiver for decoding, but unfortunately this typically requires external modulation at the transmitter to avoid excessive frequency chirp during switching. It is more effective to make use of chirp through directly modulating the current through the transmitting laser to generate *optical frequency-shift keying* (OFSK), since this avoids the need for an external modulator. The transmitted signal can then be recovered, since the frequency-shift keying of the transmitted light is translated into frequency-shift keying of the intermediate radio frequency in the receiver.

The need for two lasers at each end of the link, one for transmission and one for coherent detection, can be avoided if the same laser is used for both. The received signals at either end then become the sum of the two signals traveling in either direction, but these can be separated by electronic filtering techniques, for instance, if a subcarrier is created by a periodic variation in the transmitted optical frequency (see Figure 4.4). This gives coherent detection a potential advantage over the use of direct detection with an optical preamplifier because the pump laser for the optical preamplifier cannot be used as a transmitter.

### 4.2.3 Access Network Dimensioning

An access network using WDM could support 30 user ports with a single optical channel dedicated to each port, based on the width of the transmission window and the WDM channel spacings. If dynamic allocation of channels to ports is

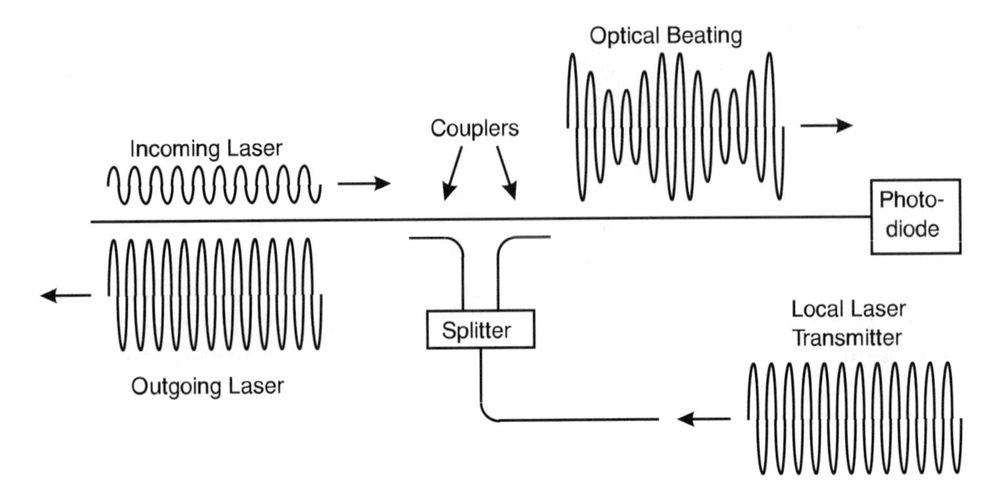


**Figure 4.4** Coherent transmitter used as local oscillator.

made, then about 300 user ports could be supported, since for certain applications, ports are often inactive for much of the time. The number of user ports could even be doubled to 600, since the same wavelength could be used by both ends of a connection, with electronic techniques being used to separate directions of transmission, assuming that most of the connections are local and could share the same fiber.

Coherent operation has the theoretical potential to increase these numbers by up to two orders of magnitude, creating an access network with the same number of user ports as a traditional narrowband exchange. This would limit the potential data rate per user, since the bandwidth per port might then be as low as 1 GHz, but this is more than adequate for foreseen communications applications. A more practical limit to reaching the bandwidth limit of the optical fiber is likely to be nonlinear effects in the fiber transmission.

## 4.3 DISTRIBUTED SWITCHING AND OPTICAL ETHER

Optical amplifiers eliminate the need for electronic bottlenecks in an optical network and extend the transmission range to hundreds of kilometers. If WDM or coherent transmission techniques are used, then the capacity of these optical networks is potentially on the order of hundreds to thousands of gigabits per second. This offers the possibility of revolutionizing the entire structure of telecommunications networks. In particular, the optical bandwidth can be viewed as a resource similar to the radio spectrum, and switching can be performed by distributed electronic nodes on the periphery of the access network. If switching is performed by wavelength selection within the access network, then the core of the telecommunications network could be reduced to intelligent gateways providing smart optical wavelength conversion. An optically transparent network that avoids the introduction of constraining electronic nodes is known as an *optical ether* (see Figure 4.5).

An analogy has sometimes been drawn between ATM and the optical ether concept. ATM is an electronic technology that aims to create a common fabric that can be used by a large number of services. The optical ether is an optical technology that also aims to create a common fabric that can be used by a number of different services. The implication, or perhaps the hype, is that optical ether will become the successor of ATM in the 21st century. However, to describe this as an oversimplification is to show remarkable constraint in the use of language, since this analogy takes no account of the level of functionality that has been specified for ATM.

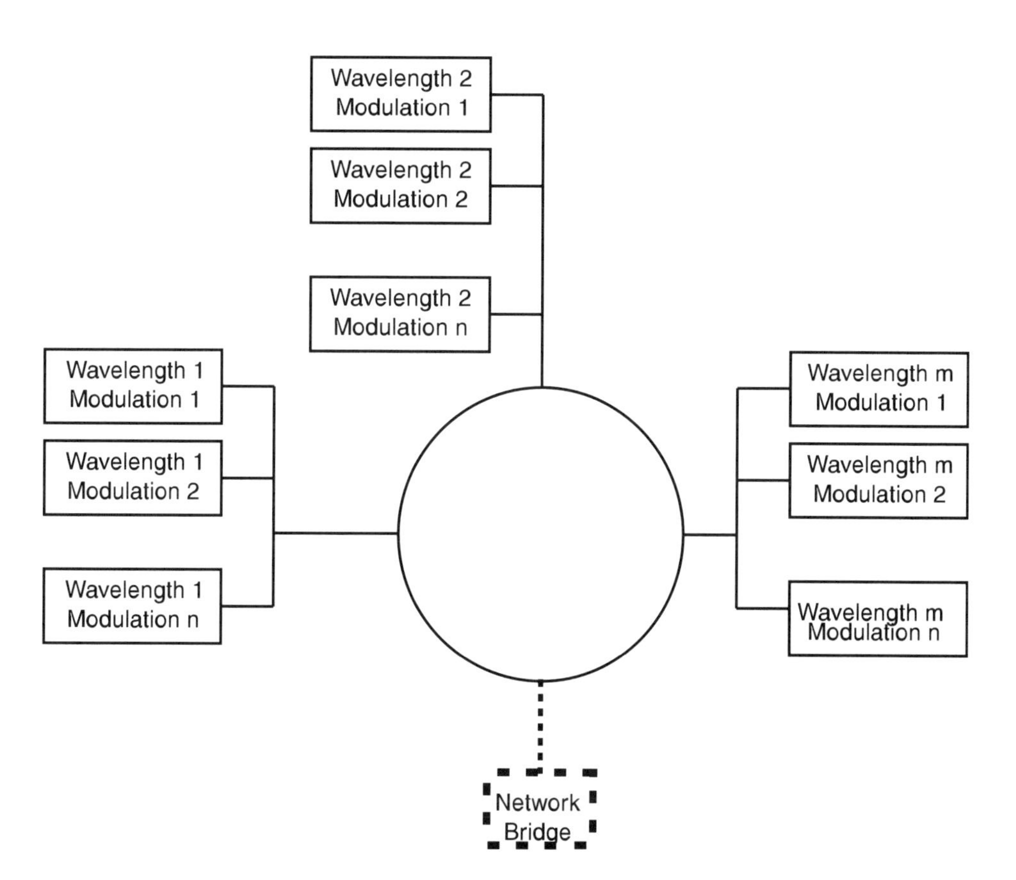


**Figure 4.5** Possible structure of an optical ether.

### 4.3.1 Optical and Electronic Modulation

The issue of spectrum allocation on the optical ether is similar to that for distributed radio systems, but has two aspects rather than one because it must cover both optical and electronic domains. The electronic domain must be included, since the absence of electronic bottlenecks within the optical network does not mean the absence of electronic modulation of the optical carriers at the network terminations. The entire optical spectrum available, typically the entire 1.55-μm window for fiber transmission, can be divided up into a number of WDM wavelengths or OFDM frequencies. An electronic multiple access technique other than TDMA can be used at each wavelength to allow the same wavelength to be shared among several network terminations. TDMA is avoided because it limits the optical transparency as it typically constrains the maximum propagation delay and because it requires a centralized source for synchronization that is not consistent with completely decentralized control.

Electronic *frequency division multiple access* (FDMA) or *code division multiple access* (CDMA) can be achieved through the binary electronic modulation of the optical carrier. The laser source can be switched on and off at different rates to generate electronic frequency modulation, and different bands of electronic frequency modulation can be allocated to different users to achieve FDMA. Alternatively, CDMA can be achieved through switching the laser source on and off in a pseudorandom sequence with different pseudorandom sequences allocated to different users. FDMA is preferable because it is easier to implement and because the softer degradation of CDMA as the number of active transmitters increases is more appropriate for nondigital, mobile communications, since a digital bit is either corrupted or not and fixed networks have less variable transmission characteristics than radio networks. CDMA is also limited by the number of appropriate pseudorandom sequences available.

The binary electronic modulation of the optical carrier is more appropriate for a WDM optical ether, since it is relatively easy to use direct detection of the optical signals from a number of users and then use electronic filtering to separate the users at different electronic subcarriers. If coherent detection of the optical signals is used in an OFDM optical ether, then the optical wavelength of the transmitters can be modulated instead of switching the transmitters on and off. This is functionally equivalent to electronic modulation of binary optical carriers, because it also allows the same optical channel to be shared between a number of users through electronic FDMA or CDMA of an optical carrier. The optical wavelength of the laser source can be modulated up and down at different frequencies, with different bands of modulation of about the same average optical wavelength being allocated to different users to give FDMA. Likewise, the optical wavelength of the laser source can be modulated according to a pseudorandom sequence to give CDMA. The advantage of OFDM is that it allows a greater number of optical channels to be used, since channel spacing is determined by electronic filtering rather than optical filtering. It also has the merit that it makes use of the intermediate electronic signal processing required in any case for FDMA.

### 4.3.2 Optical Switching

Various techniques have been proposed to replace the switching and logical operations that are commonplace in electronic systems by purely optical means. Although the potential speeds of optical systems exceed those of electronics, the delay through the optical components is often much greater than the delay through the corresponding electronic components because of the difference in their sizes. Purely optical devices are often large, particularly if they make use of nonlinear effects in optical fiber to allow one optical signal to control another, because several meters or more of fiber are required for these nonlinear effects to become significant.

These optical devices remind one of a dog walking on its two back legs. In both cases the wonder is not that it is done well, but that it can be done at all. Although optical switching is a technology that may have a significant role to play in the future, this role is unlikely to be as direct replacements for electronic switches in conventional computers or telephone exchanges, because these applications are well served by microelectronics, which itself has a significant role to play in the future. The role of purely optical switching devices is more likely to be in areas where optics is already used and there are reasons for avoiding the introduction of electronics.

Optoelectronic devices, as opposed to purely electronic devices, can be much smaller, since lithographic techniques similar to those used in microelectronics can be used with optically active semiconducting materials. As this technology develops, these devices are more likely to be used in switching applications than purely optical devices which have no electronics.

For both pure optical switching and optoelectronic switching, it is important to distinguish between data rates and switching speeds. The switching speeds for optical devices tend to be slow, whereas the data throughputs are high. This makes them more suitable for routing signals, rather than the fast switching of data or for conventional computing applications.

### 4.3.3 Establishing Connections

To establish communication between the different terminations of an optical ether, it is easier if each termination is associated with a fixed optical wavelength and electronic modulation that it constantly monitors, since this acts as a fixed address for transmission to that termination. Any other network termination can transmit to that network termination simply by transmitting at its wavelength and using its electronic modulation. Some additional sophistication is required to avoid collisions if two terminations attempt to communicate with a third, but there are a number of possible solutions to this that avoid the need for a centralized controller.

An access network based on the concept of an optical ether can have the capacity and functionality of a large switch without the need for complex centralized electronics, which would anyway act as a bottleneck for future upgrades. Having the control of the switching distributed in the terminations also makes the network robust against the changes in calling patterns likely to result from the creation of new services. The capacity of the access network can be further enhanced if only a small amount of capacity has a fixed allocation and the rest is a shared resource that may be reserved and released as required. Sharing the capacity of the optical ether potentially reduces the quality of service, but the quality can still be equivalent to that of a conventional switch, and the throughput is orders of magnitude higher.

### 4.3.4 Implementing an Optical Ether

The fixed optical filters at the receivers of the terminations of an optical ether and the encoders and decoders for simple electronic modulation pose no great technical problems for their implementation. The speed of operation for the electronics is greater than if TDMA were used because of the overhead required for the modulation, and this speed incurs the penalty of expense. The greatest problem in terms of device technology for the implementation of an optical ether is the need for inexpensive tunable lasers for the transmitters. If OFDM techniques are used, then the stability of the lasers must also be very good.

In addition to the problems of the devices required for the terminations, there are the problems of system design. In particular, the ability to upgrade the network may be compromised if all of the WDM wavelengths are already allocated. This is similar to the problem of pre-existing spectrum allocations in radio systems. The use of electronic modulation also constrains the system, since it limits the data rates at which the network terminations can communicate.

Assuming that the cost of the tunable laser transmitters is low enough, the optical ether concept may be most effective if the electronic modulation is not specified and the number of primary terminations is no more than the number of optical channels, since this imposes the least constraints on applications. This approach would not prevent a single optical channel from being shared by a number of secondary terminations in a specific application, or a single termination from using more than one optical channel.

### 4.3.5 Distributed Time Domain Switching

A less ambitious but similarly revolutionary approach to that of the optical ether, which also involves the restructuring of the telecommunications network, takes the less ambitious step of minimizing the amount of centralized electronics required. Distributed call control techniques similar to those used to establish a connection on an optical ether can also be used with little or no wavelength flexibility at the distributed nodes and with TDMA rather than FDMA or CDMA used in the electronic modulation domain.

Service transparency, rather than full optical transparency, is achieved by having the centralized controller broadcast the TDMA channels it receives to all remote terminations. This centralized controller is necessary to provide the synchronization, ranging, and leveling necessary for TDMA operation, and it is possible for the centralized controller to also provide additional functions that still leave it much simpler than a conventional switch. In particular, the centralized controller can also perform wavelength conversion functions to avoid the need for potentially expensive tunable lasers at the remote terminations.

An advantage of TDMA is that the optoelectronics can operate at lower speeds than if FDMA or CDMA were used, because there is no need to modulate an optical carrier. One of the disadvantages is that TDMA requires more sophisticated control of the dynamic allocation of time slots to remote terminations, but the same techniques that are used for more conventional TDMA systems can be used again here. The need to synchronize the remote terminations can also limit TDMA operation, but again no more so than on more conventional optical access networks.

## 4.4 THE LIMITS OF OPTICAL TRANSMISSION

Optical fiber is often discussed purely in terms of its bandwidth, ignoring the fact that it is not a perfect medium. Part of the reason for this is that the attenuation of optical fiber is low in comparison with other media and that the other operational limits are negligible if only a single optical channel of moderate data rate and signal level is used. In addition to attenuation, transmission is also limited by dispersion and by amplitude-dependent nonlinear effects.

### 4.4.1 Dispersion-Limited Transmission

Positive wavelength-dependent dispersion is conventional dispersion where high-frequency waves travel more slowly than low-frequency ones (see Figure 4.6). Negative dispersion is anomalous dispersion where the reverse occurs, for instance, at frequencies just above a resonance in the dispersive medium. Optical fibers exhibit both of these types of dispersion. The transmission over single-mode optical fiber is often limited by dispersion, so the transmission wavelength is normally selected to be close to that for the zero for dispersion.

For conventional fiber, the zero point of wavelength-dependent dispersion occurs at a wavelength close to 1.3 μm. Operation can be improved if the fiber

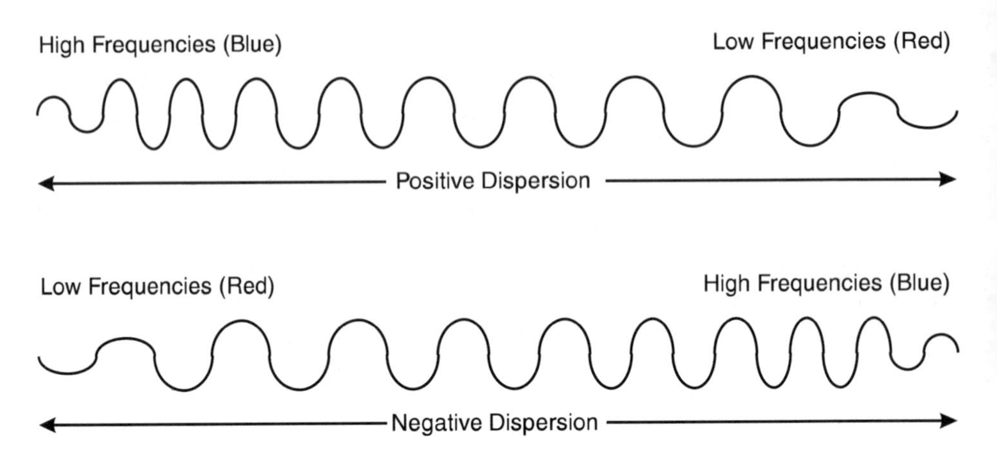


**Figure 4.6** Pulse broadening by positive and negative dispersion.

is tailored so that its waveguide characteristics shift the zero point of dispersion to about 1.55 μm, because the attenuation is even lower at this wavelength. In addition to its dependence on wavelength, dispersion in an optical fiber is also dependent on the polarization of the optical signal, but this does not have as great an effect, since the different polarization modes are randomly coupled along the fiber so that pulses are less affected. For optimum performance in the presence of dispersion, an unmodulated laser source plus an external modulator has to be used, because modulating the laser source introduces a chirp modulation of the output wavelength and a chirped pulse spreads out more because the different frequency components have different dispersions.

For dispersion-limited transmission, increasing the data rate limits the distance over which unamplified transmission can occur twice over. In addition to reducing the number of photons per bit (assuming that the transmitted power is constant), increasing the data rate also makes the pulses more sensitive to the spreading produced by the dispersion, since the pulses are smaller. This makes WDM very attractive in comparison to single-wavelength operation for dispersion-limited operation, since each WDM wavelength can operate at a lower data rate, giving an advantage that more than compensates for the lower power level at each wavelength if the total power remains the same.

### 4.4.2 Limits Due to Nonlinear Effects

In addition to the limitations due to dispersion, the use of optical fibers is also limited by a number of nonlinear effects. These nonlinearities may be due to interactions between light waves and sound waves or due to the alteration of the refractive index by electric fields.

#### *4.4.2.1 Phonon Effects*

Light traveling along a fiber can be scattered by quantized lattice vibrations known as *acoustic phonons*. The interaction between these phonons and the photons of the optical signal creates the phenomenon of *stimulated Brillouin scattering* (SBS), which permits the backward scattered signal to grow at the expense of the normal transmitted signal (see Figure 4.7). SBS limits the transmitted power levels, since it causes significant degradation of the signal if its amplitude is high enough to trigger the growth of the backward traveling wave. Dispersion-shifted fibers are almost twice as sensitive to SBS as conventional fibers.

The photons of the transmitted signal can also interact with quantized molecular vibrations, known as *optical phonons*. This creates the phenomenon of *stimulated Raman scattering* (SRS), which allows the signal photons to be shifted to lower frequencies. SRS allows interaction between the different channels in a WDM system, unlike SBS, because the frequencies of the interacting phonons are high enough to be significant in comparison with the frequencies of the photons. The coupling between WDM channels allows signals in the lower frequency channels to grow at the expense of signals in the higher frequency channels, creating crosstalk and reducing the signal levels in the higher channels (see Figure 4.8).

SRS also limits the possible transmitted power for single-channel operation, since spontaneous Raman scattering produces light that can be amplified

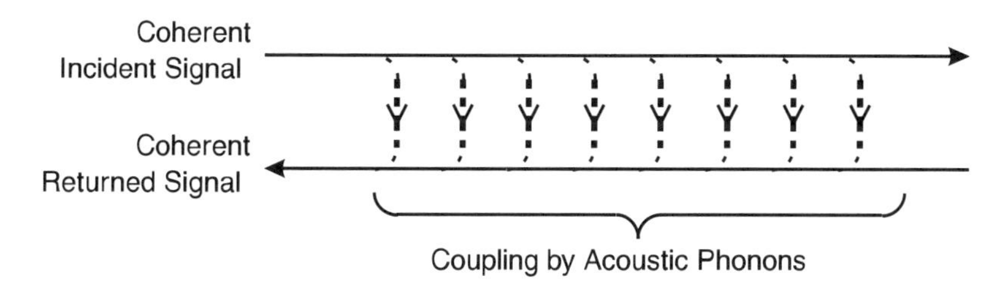


**Figure 4.7** Stimulated Brillouin scattering.

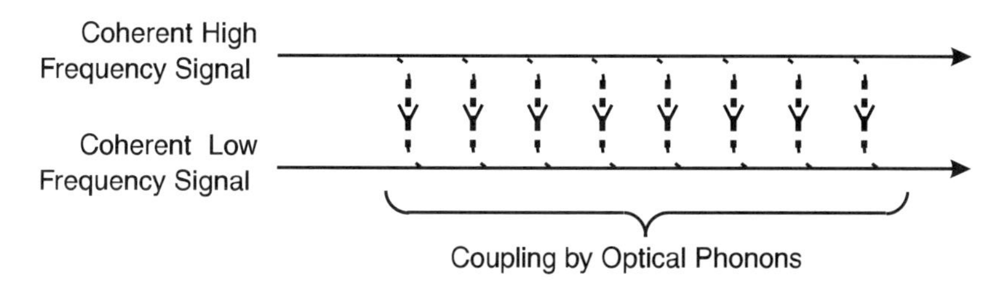


**Figure 4.8** Channel interaction due to stimulated Raman scattering.

by SRS at the expense of the transmitted signal, but the limit is higher than the corresponding SBS limit. The more significant constraint from SRS is that it sets an upper limit on the number of WDM channels. In practice, SBS does not limit the operation of real systems, because they operate at power levels at which SBS is not significant. However, SRS does impose a practical limit for real systems because it is severe enough to create a trade-off between the number of channels and the transmission distance. An access network may be able to operate with about 100 channels over about 20 km, but the larger the catchment area, the smaller the number of channels.

#### 4.4.2.2 Refractive Index Effects

The broadening of optical pulses due to positive dispersion is aggravated at high optical signal levels, because the refractive index of optical fiber increases with the intensity of the signal being transmitted. This increase of refractive index with intensity, known as the *Kerr effect*, occurs because intense light increases the polarization of the medium by distorting the orbits of electrons, and this slows the transmission of the light itself. The optical waveform is slowed more strongly near the center of the pulse, where the intensity is greatest, distorting the phase of the waveform in a phenomenon called *self-phase modulation* (SPM) (see Figure 4.9).

The result of the Kerr effect is that the tail of the waveform is compressed in time and the leading edge becomes elongated as the high-intensity center of the waveform is slowed down. This increases the spectrum of the pulse with the tail of the pulse more blue and the leading edge of the pulse more red. If the medium has positive dispersion at the pulse frequencies, the pulse is then broadened by the differential dispersion between the frequencies of its edges. The phase distortion from SPM accumulates as the path length increases, and becomes significant for long transmission lines.

If several optical signals at different wavelengths travel down the same fiber, then the increase of refractive index with optical intensity results in the signal in one channel modulating the phase of the signal in other channels. This phenomenon is known as *cross-phase modulation* (XPM). The frequency broad-

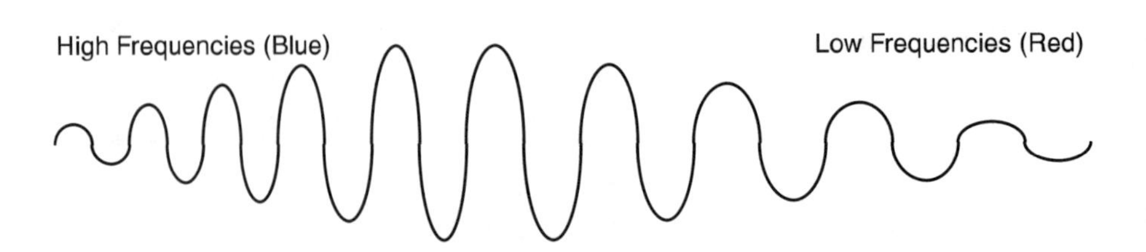


**Figure 4.9** Self-phase modulation due to the Kerr effect.

ening corresponding to this phase modulation is reduced because the dispersion of the fiber causes pulses in different channels to pass through one another, producing upshifts and downshifts in frequency that tend to cancel out.

The beating together of optical signals at two different frequencies also produces a modulation of the refractive index of the fiber at a frequency corresponding to the difference between the two original frequencies. This modulation interacts with the original signals to produce sidebands that are offset by the difference in the original frequencies. The higher sideband of the lower frequency merges with the higher frequency and vice versa, leaving one sideband below the lower original frequency and one sideband above the higher original frequency (see Figure 4.10). Because the result is a set of four waves, it is known as *four-wave mixing* (FWM) or *four-photon mixing* (FPM).

The generated sidebands grow at the expense of the original signals, reducing their strength, and the mixing creates crosstalk between them. Dispersion significantly reduces the effect, since it causes the modulation to travel at a different speed from the original signals, and the reduction in the mixing is greater if the original signals are well separated. Near the zero point of differential dispersion, FWM can significantly limit how closely WDM channels can be spaced. FWD can also be a major limitation if the WDM channels are equally spaced, because sideband channels then coincide with other operating channels.

## 4.5 SOLITON TECHNIQUES

The 35-nm optical window at around 1.55 μm corresponds to a bandwidth of about 4,000 GHz, and it has been suggested that this could be completely used for communication. Even if optical amplifiers are used to compensate for attenuation and splitting losses, dispersion limits the achievable data rate at a single wavelength to under 20 Gbps for the line lengths required for an access network. It is not feasible to make use of this bandwidth by using many different operating wavelengths, because even if coherent OFDM techniques were used instead of WDM (which has a limited number of channels because of the bandwidth of the optical filters), nonlinear effects would be significant because of the high total optical power in the fiber. Soliton transmission may

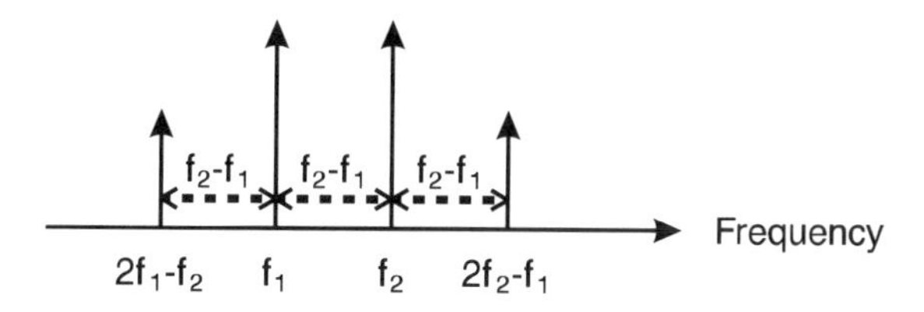


**Figure 4.10** Four-wave mixing.

be able to make better use of the potential of optical fiber, because solitons can have pulse widths that are smaller than conventional pulses, since solitons are not limited in the same way by dispersion.

### 4.5.1 Soliton Transmission

An optical pulse cannot consist of a single optical wavelength, because of the Heisenberg uncertainty principle. For simple laser diodes, pulses consist of an even greater spread of wavelengths, because lasers chirp in frequency when they are switched on and off. Any normal optical pulse will spread out over time, because its different wavelength components travel at different speeds. This effect becomes more pronounced with distance, and on high-speed trans-atlantic cables it has been necessary to detect the broadened pulses and regenerate the original short pulses to compensate for this dispersion. In optical access networks, the effect is less because the distances are shorter, but even over these shorter distances the effect limits the ultimate transmission rates, since small amounts of pulse broadening become significant if the pulse widths are also small.

Normally nonlinear SPM aggravates the dispersion of the pulse, because positive dispersion and SPM cooperate to broaden the pulses. A short, intense pulse injected into a fiber is modified by SPM so that the red (low-frequency) components accumulate at the leading edge and the blue (high-frequency) components accumulate at the trailing edge. If the transmission occurs in the negative dispersion region, then the blue components at the trailing edge move faster than the red components at the leading edge, compressing the pulse and increasing the SPM, which compensates by driving the red components back to the leading edge and the blue components back to the trailing edge (see Figure 4.11).

An optical pulse that does not disperse over time because it is balanced in this way is known as a *soliton* (i.e., a solitary wave similar to a diffuse

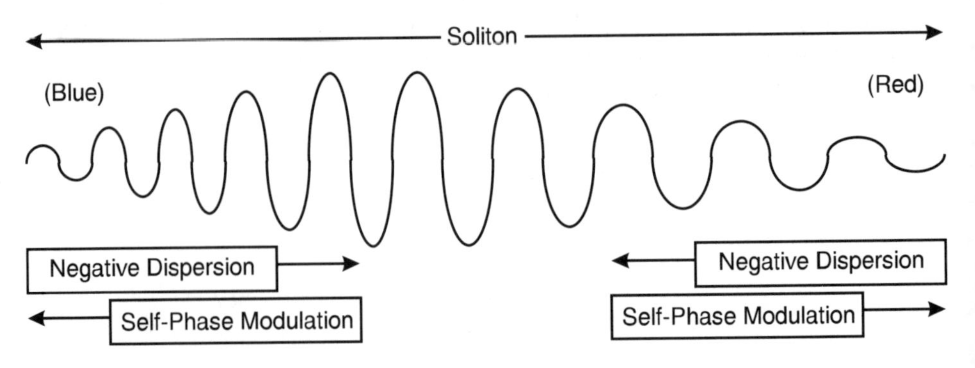


**Figure 4.11** Dynamic balancing of negative dispersion and self-phase modulation.

particle). It turns out that in many cases the balancing of the two effects produces a stable equilibrium and the resulting solitons are robust entities. If the transmission is not amplified, then the soliton eventually dissipates, since it is attenuated by propagation, causing the SPM to decrease until the soliton is unable to keep the red and blue frequency components apart.

### 4.5.2 Soliton Interactions

Solitons attract or repel each other depending on their relative phases (see Figures 4.12 and 4.13). A single soliton maintains its form through the counteraction of negative dispersion by SPM. If two solitons approach each other, then the waveforms will either enhance each other or cancel each other out, depending on their relative phases. If the waveforms enhance each other, then the increased SPM between the solitons forces the adjacent frequency components to the far sides, producing attraction. If the waveforms cancel, then the solitons repel, since the reduced SPM between them allows the negative dispersion to move the remote edges farther apart.

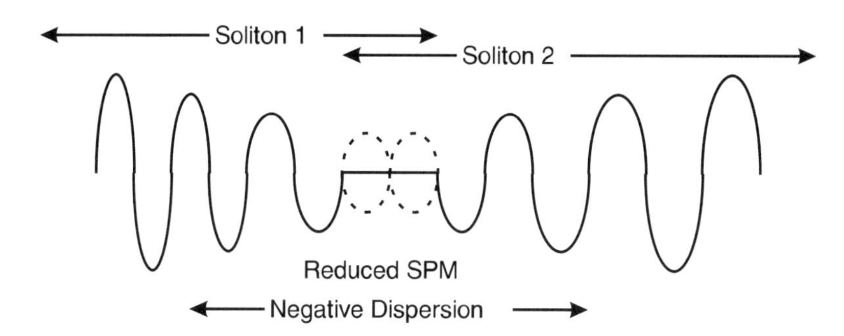


**Figure 4.12** Solitons out of phase repelling.

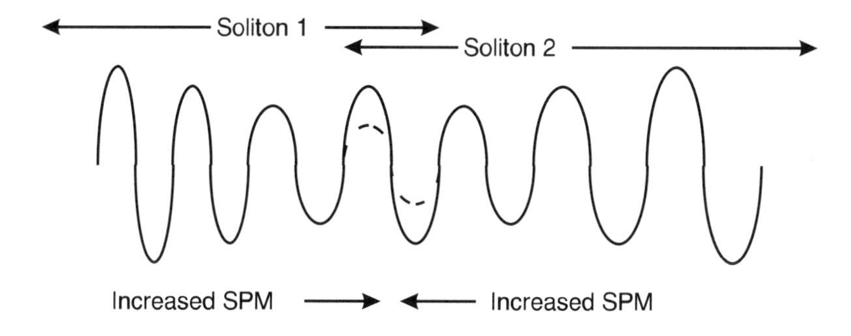


**Figure 4.13** Solitons in phase attracting.

If two solitons that attract also have the same speed and are brought together, then a new soliton, which is the bound state of the two original solitons, can be created. By repeating the process, solitons that are the bound state of several fundamental solitons can be generated.

Solitons that are a large distance apart can also interact through an opto-electronic effect. Solitons are accompanied by acoustic shock waves that are reflected by the wall of the fiber and can affect a soliton some distance away through the piezoelectric effect. These shock waves are reminiscent of the Cherenkov radiation, which occurs when a charged particle travels faster than the speed of light within a gas or material, and the interaction is reminiscent of the acoustic interaction between separated pairs of electrons, which is the source of superconductivity.

### 4.5.3 Limiting Factors for Soliton Transmission

If optical amplifiers are used for soliton transmission, then the amplifier noise combines with the solitons through nonlinear interactions. These interactions have components that shift both the position and the frequency of the solitons. Although the direct shifts in position are negligible, the shifts in frequency cause indirect shifts in position that grow with time due to dispersion, and the resultant jitter in the position of the soliton pulses is significant. This jitter produces the Gordon-Haus limit for soliton transmission.

The Gordon-Haus limit can be overcome by introducing a number of optical filters along the transmission path, which herd the solitons back to the correct frequency. For best results, the peaks of the optical filters need to be offset slightly from the peaks of the adjacent filters so that almost all of the amplifier noise is blocked while the nonlinear solitons can slither through the filters. Using filters in this way restricts the ability of the transmission system to support conventional WDM, since the net effect of the staggered filters is a very narrow filter end to end or even an opaque medium. Solitons do not find the medium opaque fiber, because they shift in frequency as they pass through the filters. To obtain WDM operation with solitons, Fabry-Perot filters may be needed to overcome the Gordon-Haus limit, since these filters have multiple peaks that can be used to create an opaque path for herding solitons and a transparent path for WDM operation.

Even if amplifiers are not used, filters may still be needed to overcome the effect of interactions between solitons at different wavelengths. Solitons at different wavelengths travel at different speeds because of dispersion and attract and repel each other as they pass through each other. If the interaction was constant, then it could be ignored, because it would create a constant offset. However, the interaction is not constant, because the data carried is continually changing and this has a similar effect as amplifier noise, producing shifts in position and frequency. Conventional optical transmission may be able to support about 30 WDM transmission channels, but fewer channels may be possible if soliton transmission is used.

There are trade-offs between conventional and soliton transmission systems, because conventional transmission systems can operate with greater distances between amplifiers, since there is no need to maintain the intensity of the pulse to ensure adequate SPM. This has a greater impact on long-haul applications than for the access network, because fewer stages of amplification may be necessary in the access network. If a large number of lower rate channels are required, then conventional optical transmission may have an advantage, but soliton transmission may be necessary if the total required throughput becomes extremely high, since conventional operation on a large number of channels is also limited by nonlinear effects.

### 4.5.4 Solitons in the Access Network

Like the deployment of optical amplifiers, the deployment of soliton transmission systems in the access network is likely to lag behind the deployment of soliton transmission systems on transoceanic routes, because the immediate demand for high-capacity transmission in the access network is less. Once the technology has matured and is deployed for transoceanic and other long-haul applications, then it may also be deployed in access networks.

Soliton transmission systems may have an advantage in access network applications because the distances are shorter. This means that there is less need for optical amplifiers to offset the attenuation and maintain the pulse intensity required for soliton transmission. This advantage is mostly relevant to simple star topologies, because they only suffer from normal fiber attenuation.

If PON-type passive multistar topologies are used, then soliton transmission is compromised, because significant additional attenuation is produced by the passive splitters. For these topologies, an optical amplifier is required before the splitter, or even integrated with it, to prevent the splitter loss from reducing the pulse intensity to a level where soliton operation is not possible.

## 4.6 QUANTUM CRYPTOGRAPHY

The simplest types of cryptography rely on the use of a secret algorithm to transform a message into ciphertext. The security of these schemes relies on the security of the secret algorithm, and can be compromised if even the general type of the algorithm is known, especially if some general information about the nature of the message is known. For example, most codes that substitute a certain letter or character for another can be broken if the language of the message can be guessed, because of the frequency of occurrence of letters and combinations of letters. This approach was once described by Sherlock Holmes (in "The Adventure of the Dancing Men" from *The Return of Sherlock Holmes*).

### 4.6.1 Cryptographic Keys

More sophisticated types of coding use both an algorithm and a key, since the number of different coding algorithms is limited. For example, the letters of the alphabet can be represented by different characters according to the key. This approach is not completely secure if the key is shorter than the message, since for short keys it is possible to guess the key length and treat the problem as several related problems where no key is used. However, for more sophisticated algorithms, a key of a reasonable length can provide a level of protection it has not been possible to break.

This use of keys creates the problem of finding a secure way to distribute these keys, since a different key is required for each pair wishing to communicate. One approach is to have an algorithm that has two different keys, one for encryption and one for decryption. Although in theory the key for decryption can be deduced from the encryption key if the algorithm is known, this is not possible in practice if a mathematically intractable puzzle, such as the factorization of certain types of large numbers, needs to be solved. The encryption key can then be made public so that anyone wishing to send a secure message to the owner of the key can use the key to encrypt the message, but only the owner who knows the decryption key can decrypt it.

Unfortunately, the algorithms for public key techniques are typically slow because a mathematical puzzle that is impossible in practice to solve often takes a computer a significant length of time to create. This can make public key techniques unsuitable for the encryption of real-time communications, but is not a significant constraint if it is used to provide secure communication for the keys for other encryption algorithms. This use of public key cryptography should allow secure communications to be bootstrapped over an insecure link, but the effectiveness of the process depends on the absence of technological innovations or mathematical discoveries that make solving the mathematical puzzle practical. There is also the problem that eavesdropping can still be used on this or other techniques intended for secure communications, allowing valid messages to be recorded or duplicated.

### 4.6.2 Quantum Cryptography

In a universe that obeyed the laws of classical physics, it would always be possible to devise a way to tap a signal that could not be detected by a known security device, since the disturbance created by the tapping could be made less than what the security device could detect. In reality, the universe obeys laws of quantum physics, which make it impossible to make certain measurements in a way that does not disturb the system in a detectable manner. The Heisenberg uncertainty principle means that no system can have exact values for certain pairs of noncommutating observables, such as momentum and position or spin along different axes. The measurement of one observable of a pair ensures that information about the other observable is lost.

In recent years experiments have been carried out that make it clear this is not simply a matter of the measurement of one observable causing information about the other observable to be lost. Information is also lost if it is possible for the measurement to be made, even if it has not actually been made. In addition, experiments demonstrate that information in the present is still lost even if the measurement of the other observable is delayed to some time in the future. Although it has not been possible to use this effect to transmit information from the future to the present, it is possible to use it to detect the unauthorized tapping of a normal transmitted signal.

If you find these ideas difficult to believe, then you are in good company. Einstein was uncomfortable with these aspects of quantum theory, which have now been experimentally confirmed. Richard Feynman, who has been described as the greatest theoretical physicist of the mid-twentieth century, said that anyone who thought that they understood them was mistaken. The only people who have seemed happy about the ideas are those lecturers in physics who for years have dismissed students who have questioned them as failing to understand the course work.

The polarizations of a photon along nonorthogonal axes correspond to noncommutating observables, since they are components of its spin. Data can be encoded so that a 1 corresponds to a vertical polarization and a 0 corresponds to a horizontal polarization, or alternatively data can be encoded along axes that are at 45 deg to the horizontal and vertical (see Figure 4.14). If an eavesdropper measures the polarization along an axis that is at 45 deg from that along which it was transmitted, then the polarization is altered in a way that can be detected by the genuine receiver.

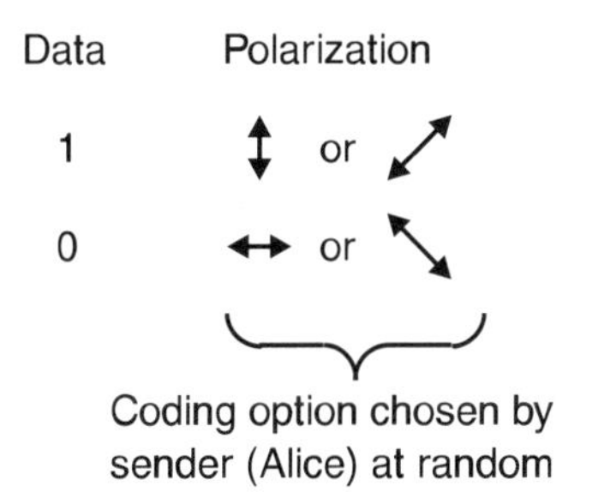


**Figure 4.14** Encoding of data for transmission.

The transmitter and the receiver can independently choose at random between the two sets of axes for encoding and sensing the data, and then discard those cases where they have chosen different orientations (see Figure 4.15). If an eavesdropper attempts to measure the polarization and retransmit a photon with the measured polarization, then this introduces a 25% error rate in the transmission system, because there is a 50% probability of the eavesdropper choosing the wrong axis, in which case there is a 50% probability of the genuine receiver measuring the wrong result (see Figure 4.16).

In theory, similar techniques can be applied to ensure there is no unauthorized reading of stored information before it is used, and this can be combined with the distribution of information to different locations. Instead of a single polarized photon being transmitted, two coordinated photons that have opposite polarizations can be used. These photons are then stored in different

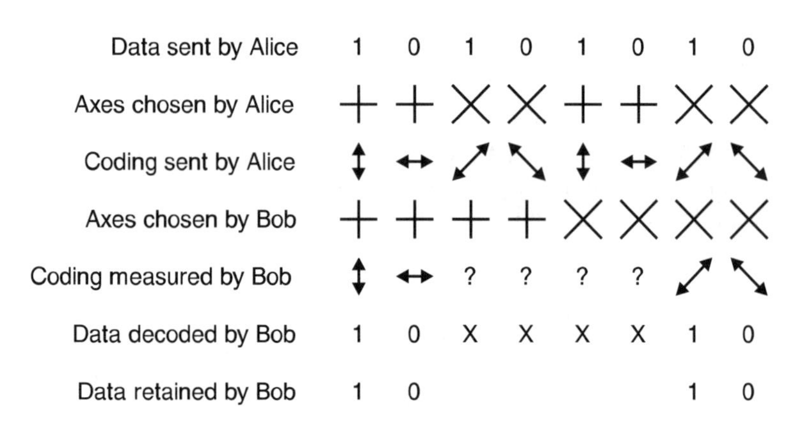


**Figure 4.15** Normal transmission from Alice to Bob.

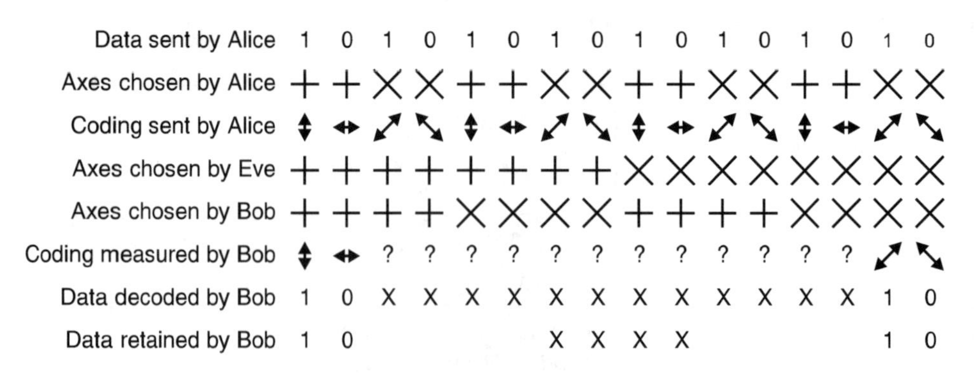


**Figure 4.16** Transmission with eavesdropping by Eve. Xs in data retained by Bob imply 25% error rate due to eavesdropping by Eve.

locations until they are needed. According to the *Einstein-Podolsky-Rosen* (EPR) effect, the measurement of the polarization of one photon of a pair in one location affects the polarization of the other photon in the other location. This is a counterintuitive quantum effect resembling action at a distance with no physical transmission between the locations at the time of measurement. Any unauthorized measurement of the stored photons can be detected, because it causes their waveform to collapse into a specific state, altering the behavior of the photons. Unfortunately, although this is a fascinating physical phenomenon, it is not a practical means of storing information securely, since photons cannot be stored for more than a small fraction of a second.

### 4.6.3 Access Network Implementations

In a practical implementation of quantum cryptography, it is important to prevent more than one photon being transmitted per bit, because this would allow a photon to be tapped and measured without disturbing the photon detected by the genuine receiver. This is achieved by attenuating the transmitted output so that the average number of photons per bit is much less than one. Although this means that additional results need to be discarded because no photon is detected by the genuine receiver, it makes the probability of both an eavesdropper and the genuine receiver simultaneously receiving photons so low as to be negligible.

Since the use of polarizations as complementary variables for quantum cryptography relies on distinguishing between polarization orientations at the receiver, the optical fiber must not randomly couple the polarization modes along its length. The random coupling of polarization modes along the fiber is normally an advantage because it ensures that polarization-dependent dispersion can be ignored, since it is smaller than the frequency-dependent dispersion. There will be additional jitter on an access network that supports quantum cryptography, because the polarization modes have not been randomized.

Quantum cryptography can also be implemented on an access network with a passive multistar topology, since individual photons cannot be split. The effect of the passive splitters is to increase the attenuation on any one link and to distribute the photons among the remote ends at random so that the system behaves like a number of quantum cryptography channels in parallel. Care may again need to be taken that the fiber splicings do not randomize the polarization orientations, because this would disrupt operation. Polarization orientations may also have to be synchronized between the transmitter and the receiver before transmission to take account of environmental changes.

Avalanche photodiodes may need to be used in the receivers, since they need to be able to detect individual photons. The detectors will also suffer from dark current, which will result in errors that cannot be differentiated from eavesdropping. This makes it impossible to be certain that no eavesdropping has occurred, only that the level of eavesdropping is extremely unlikely to be above a value calculated from the error rate. This makes it necessary to use more conventional techniques to ensure that communication is secure even in the presence of eavesdropping that is guaranteed to be below a certain level.

Various well-known error detection techniques can be used to ensure that data corrupted by eavesdropping or noise are ignored. The number of detected errors allows the upper limit on the amount of eavesdropping to be calculated, since the implementation determines the percentage of the eavesdropped bits resulting in errors. The uncorrupted bits, for which the probability of eavesdropping has been calculated, can then be used to select a subset of themselves. If the upper limit on the amount of eavesdropping is low, an eavesdropper has great difficulty in knowing which subset has been chosen, since the eavesdropper does not know most of the information necessary to work this out. The details of this technique, known as *privacy amplification*, can be chosen so that for detected errors below a certain threshold, the probability of an eavesdropper being able to discover even one bit of the selected information is extremely low.

## 4.7 SUMMARY

Optical amplifiers are a key component in many advanced optical systems. They operate by coupling the light from a pump laser into a loop of erbium-doped fiber that is added to the normal optical fiber. An incident optical signal stimulates the decay of the excited erbium ions, amplifying the signal.

Optical amplifiers are very effective when used for preamplification of an optical signal before it is detected by a low-noise PIN photodiode. They are also simpler and more robust than regenerating optoelectronic repeaters and do not create electronic bottlenecks which limit future upgrades. Unlike repeaters, optical amplifiers generate a certain amount of optical noise, which constrains their maximum practical gain. This generated noise also means that optical amplifiers need to be more closely spaced than repeaters and makes optical filtering before signal detection more important.

A PIN photodiode with optical preamplification is almost as sensitive as a coherent optical detector, but coherent systems make better use of the window for optical transmission, since WDM channels would need to operate at extremely high data rates to fill this window. Coherent systems support a larger number of more closely spaced channels than WDM systems, because they perform channel separation electronically on the beat signal between the incoming optical signal and the local laser oscillator. Coherent transmission can make use of the chirping of laser diodes to create OFSK, and it is theoretically possible to also use the transmitting laser as the local oscillator for coherent detection.

Coherent operation would allow an access network to support as many ports as a narrowband local exchange, but with 1 GHz of bandwidth possible at each port. Unfortunately, in addition to being less robust than direct detection systems, coherent systems also require more expensive lasers, since their lasers need to have greater frequency stability.

The optical ether concept is that of a decentralized, optically transparent network that avoids constraining electronic nodes so as to make the best use of the window for optical transmission, and offers the possibility of completely restructuring telecommunications networks. Transmission on an optical ether is optically multiplexed, using WDM or coherent techniques, and electronically multiplexed, preferably using FDMA, since CDMA is less effective and TDMA is not decentralized. Connections can be established using a decentralized call control. However, pure optical switching techniques seem unlikely to replace electronic switching, since they are better suited to the routing of optical pathways rather than the switching of bits, cells, or packets, although it may be easier to make use of hybrid optoelectronic switches.

A less ambitious approach than a full optical ether, but one that is still revolutionary, is to allow a small amount of centralized control to support the synchronization, ranging, and leveling functions required for TDMA operation. This may be complemented by centralized wavelength conversion and gateway functions, but again the call control can be decentralized.

Optical-fiber transmission is often limited by differential dispersion, which causes pulses to spread because the different frequency components of the pulses travel at different speeds. Fibers are often designed so that the dispersion due to their waveguide characteristics shift the zero-point of differential dispersion to about 1.55 μm, where the attenuation has a minimum. As the optical signal intensity increases, the light interacts with quantized lattice vibrations (SBS) and molecular vibrations (SRS) to amplify either backward or forward traveling waves. In the worst case, the resulting crosstalk from SRS produces a trade-off between the transmission distance and the number of transmitted channels. The refractive index is also increased locally when the optical signal strength is high, producing a rainbow-like spectrum separation for the pulse (SPM), which can exaggerate the pulse spreading due to dispersion. This can also produce frequency shifts due to interactions of pulses in different channels (XPM) and mixing between optical channels (FWM or FPM), which limits how closely optical channels can be spaced.

Solitons are very short and intense optical pulses which are not spread out by dispersion because they exist at optical frequencies where the dispersion is counteracted by SPM, producing a stable waveform that resembles a particle. Solitons need to be boosted by optical amplifiers, since otherwise they are attenuated until they behave like other pulses. The transmission of solitons is limited by their interactions with amplifier noise (the Gordon-Haus limit), and although this can be ameliorated by appropriate filters, these filters restrict conventional forms of transmission. Soliton transmission is also limited by the attraction and repulsion between solitons in adjacent channels and this may prevent an optical fiber from supporting as many channels for soliton transmission as it could for WDM operation. Point-to-point soliton transmission in access networks may be easier than on long-haul routes, because optical amplification may be unnecessary, but amplifiers will again be needed if there is passive splitting within the access network.

Conventional approaches to secure communications rely on the impracticality of decoding an intercepted message that has been encrypted, often using a key in addition to an encryption algorithm, because ignorance of the encryption algorithm is not a safe assumption, since the number of different encryption algorithms devised is limited. The laws of classical physics allow the construction of very sensitive devices to eavesdrop on transmissions with negligible interference, but the laws of quantum physics prevent the undetectable eavesdropping of certain forms of optical transmission. Secure optical-fiber transmission using a quantum effect (quantum cryptography) has been demonstrated and can be adapted for use on access networks with passive splitting, although care must be taken with polarization couplings. The technique of privacy amplification is also needed, since it can be difficult to distinguish between low levels of eavesdropping and naturally occurring errors. Privacy amplification can ensure that the probability of successfully eavesdropping on even a single bit of data is extremely low if the maximum amount of eavesdropping can be measured.

## Selected Bibliography

Baack, C., and G. Walfi, "Photonics in Future Telecommunications," *Proc. IEEE*, Vol. 81, No. 11, November 1993, pp. 1624–1632.

Bennett, C. H., G. Brassard, and A. K. Ekert, "Quantum Cryptography," *Scientific American*, October 1992, pp. 26–33.

Cochrane, P., and M. Brain, "Future Optical Fiber Transmission Technology and Networks," *IEEE Communications Magazine*, Vol. 26, No. 11, November 1988, pp. 45–60.

Cochrane, P., D. J. T. Heatley, P. P. Smyth, and I. D. Pearson, "Optical Telecommunications—Future Prospects," *Electronics & Communication Engineering J.*, Vol. 5, No. 4, August 1993, pp. 221–232.

Gidron, R., and A. S. Acampora, "A User Tunable Access Lightwave Network," *J. Lightwave Technology*, Vol. 11, No. 5-6, May–June 1993, pp. 971–978.

Haus, H. A., "Optical Fiber Solitons, Their Properties and Uses," *Proc. IEEE*, Vol. 81, No. 7, July 1993, pp. 970–983.

Khoe, G.-D., "Coherent Multicarrier Lightwave Technology for Flexible Capacity Networks," *IEEE Communications Magazine*, Vol. 32, No. 3, March 1994, pp. 22–33.

Li, T., "The Impact of Optical Amplifiers on Long-Distance Lightwave Telecommunications," *Proc. IEEE*, Vol. 81, No. 11, November 1993, pp. 1568–1579.

Miki, T., "Optical Transport Networks," *Proc. IEEE*, Vol. 81, No. 11, November 1993, pp. 1594–1609.

Nakazwa, M., "Soliton Transmission in Telecommunication Networks," *IEEE Communications Magazine*, Vol. 32, No. 3, March 1994, pp. 34–41.

Tonguz, O. K., and K. A. Falcone, "Gigabits-Per-Second Optical Interconnection Networks: Fault-Tolerance With and Without Optical Amplifiers," *J. Lightwave Technology*, Vol. 12, No. 2, February 1994, pp. 237–246.

Townsend, P. D., "Secure Key Distribution System Based on Quantum Cryptography," *Electronics Letters*, Vol. 30, No. 10, 12 May 1994, pp. 809–811.

Townsend, P. D., S. J. D. Phoenix, K. J. Blow, and S. M. Barnett, "Design of Quantum Cryptography Systems for Passive Optical Networks," *Electronics Letters*, Vol. 30, No. 22, 27 October 1994, pp. 1875–1877.

van Tilborg, and C. A. Henk, *An Introduction to Cryptology,* Boston: Kluwer Academic Publishers, 1988.

# Radio Access Technology 5

"Radio Ga Ga"

—Queen

This chapter describes the radio systems that may form part of an access network linking customers to the core network. These systems may provide the complete link or they may be used in conjunction with copper and optical-fiber transmission systems. Satellite communication systems will not be included here because they are not normally considered to be part of an access network.

Both mobile (cellular) communications systems and cordless systems are discussed, since this helps to clarify the relationships between them, but cordless systems may have a greater impact on access networks, since this technology is less expensive and has immediate applications. Although neither *local multipoint distribution service* (LMDS) at 28 GHz nor *licensed millimeter-wave service* (LMWS) at 48 GHz is discussed here, both should be noted for the future, since they could provide competition for cabled video and telephony services, at least in North America.

## 5.1 BACKGROUND

Radio links have been used for many years to connect customers to the telecommunications network, particularly when there have been difficulties using a traditional copper link. Sometimes radio links have been used for expediency, especially for business customers. In the case of expedient deployment, the intention has been for the link to be replaced when it becomes feasible to use a copper link instead, and the radio equipment is then recoverable for expedient use elsewhere.

Closely related to this expedient use of a radio link is the permanent use of the radio equipment when it is not feasible to replace it with copper. This can occur with remote or isolated customers who must be provided with service, because it is necessary to do so under the terms of universal service obligation, which may form part of the operator's license. Typically the radio channels are dedicated for the use of the customer in a point-to-point transmission configuration, and the signals may need to be transmitted with considerable power to achieve the necessary range.

Recent developments in mobile and cordless radio systems are making radio links more attractive and creating the possibility of their use as a permanent alternative to copper in both rural and urban or suburban areas. In these applications, several customers within a small geographic cell are allocated radio bandwidth on demand in a point-to-multipoint configuration. The radio power can be kept low because the distances are small and the radio spectrum can be reused in other cells.

## 5.2 RECENT DEVELOPMENTS IN RADIO SYSTEMS

By the early 1990s, Europe had outstripped the United States and Japan in the development of second-generation cellular and cordless telecommunications. Ironically, part of the reason for this was the success of the first-generation systems outside of Europe.

Both cellular and cordless systems are based on operation over short distances. The power of the radio transmitters is kept low to limit the range of operation, and the frequencies can be reused in other physical locations. This is similar to the situation at a party where people are only able to talk to those who are near them, but many conversations are possible in different parts of the room. In contrast, the traditional use of radio is rather like having the music turned up so high that normal conversation is not possible.

Cellular radio is able to support mobile operation by dividing an area into geographic cells, each served by a base station (see Figure 5.1). Communication is maintained while traveling from one cell to the next by intercell handoff of the telephone connection between the base stations of the adjacent cells. The telephone connection is handed over from the cell that is being left to the cell that is being entered. The common base station of each cell handles all of the calls within the cell using point-to-multipoint operation.

Cordless transmission in first-generation equipment was restricted to short distance, point-to-point operation from a handset to its dedicated base station. However, the distinction between cellular and cordless started to blur in second-generation systems with the introduction of point-to-multipoint operation for cordless transmission, allowing one base station to serve a number of handsets within its cell. The distinction became further blurred by allowing a handset to work with a number of different base stations. Finally, the distinction became almost completely lost when handoff was added to allow mobility

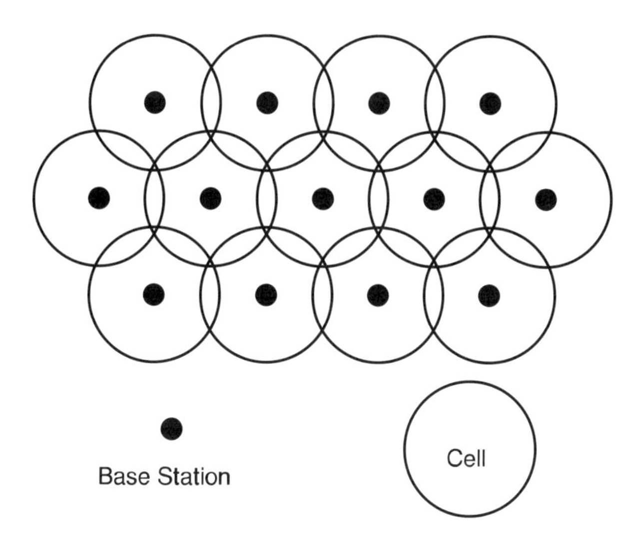


**Figure 5.1** Division of a region into overlapping cells.

between cells, although this was initially limited to cells at a single business location.

Standards for first-generation cellular systems were rapidly developed in the United States and in Japan (see Table 5.1). In the United States and Canada, the *advanced mobile phone service* (AMPS) standard was adopted in the early 1980s. A similar system was adopted in Japan using 25-kHz spacings instead of the 30-kHz adopted in the United States. It would surely be wrong to believe that this difference was agreed on to help protect the domestic Japanese market from foreign imports, even if that is a consequence.

The proliferation of national specifications within Europe prevented a single first-generation mobile telephone from being used throughout Europe. To overcome this limitation, the *Groupe Speciale Mobile* (GSM) specification was rapidly agreed to as a second-generation pan-European standard. Likewise, the threat of a wide variety of imported cordless systems operating at a variety of different frequencies and power levels led to the adoption of the CEPT first-generation *cordless telephony* (CT1) standard. However, in France and in the United Kingdom, the CT1 standard was not adopted because it was thought to be too expensive and limited to business users, and both countries created national standards.

In the United Kingdom, the government's Department of Trade and Industry sponsored a second-generation specification for cordless operation, CT2, to which a specification for a *common air interface* (CAI) was added in the late 1980s. This was endorsed as an interim ETSI standard by the end of 1991, but is not mandated by a European Community (EC) directive and so need not be adopted by all countries in the EC.

A more sophisticated second-generation cordless standard, the *digital European cordless telecommunications* (DECT) standard, was created because it was impossible to reconcile differences between supporters of FDMA and of TDMA, so the politically correct DECT standard uses both. This political correctness is acknowledged, since the standard is mandated by an EC directive, and countries in the EC are required to make bandwidth available for it near 1.8 MHz.

Europe has also adopted a second-generation cellular radio specification for *personal communications networks* (PCN), which have a greater emphasis on services. European operators preferred to base this on GSM rather than DECT. This is the *Digital Communications System 1800* (DSC1800) standard, with the name reflecting that its GSM basis has been modified for operation at 1,800 MHz.

Second-generation developments in the United States and Japan were slower in the 1980s. Eventually a second-generation cellular specification was agreed on in the United States. This is the Electronic Industry Association Interim Standard, IS-54, which is designed to coexist with the first-generation AMPS standard and to use the same channel spacings. The same considerations apply to the Japanese second-generation standard and the Japanese frequency spacings.

**Table 5.1**
Cellular and Cordless Radio Standards

| *First Generation* | | *Second Generation* | |
|---|---|---|---|
| Cordless | Cellular | Cordless | Cellular |
| CT1 (Europe—but national standards in France and U.K.) | AMPS (North America—30 kHz spaced) | CT2 (Europe) | GSM (Europe) |
| | (Japan—25 kHz spaced) | DECT (Europe) | DSC1800 (Europe) |
| | | | IS-54 (North America) |

## 5.3 CELLULAR AND CORDLESS TECHNOLOGY

The original goal of cellular radio technology was to support car phones, which could be fast moving and a considerable distance from the base stations of their cells. In contrast, cordless telephony was developed to operate over shorter distances with stationary or slowly moving terminals.

There are greater restrictions on the bandwidth for cellular telephony because the larger transmission distances mean that the transmitters are more powerful and so potentially more disruptive to other types of equipment.

Cellular telephony must also be able to operate with terminals whose location may be initially unknown and over channels with rapidly changing characteristics. In addition to all this, the number of terminals within a cell may also change rapidly due to road traffic conditions.

### 5.3.1 First-Generation Systems

First-generation systems are characterized by the use of analog frequency modulation to support the voice traffic and the use of separate transmit and receive frequencies. They typically also adopt a rather crude approach for control and signaling.

First-generation cordless systems were designed to operate only with their own private base station, often using a proprietary protocol. Most only use fixed frequency channels, although the more sophisticated allow manual channel frequency selection by the user. This channel selection can be a useful feature, because the systems are prone to interference from neighboring users who can operate on the same frequency. First-generation cordless systems have no means of coordinating the use of radio spectrum between users. They were also designed for use within a small building and so have a range of a few tens of meters.

Cellular systems require more elaborate signaling than cordless because they must be able to transfer ongoing calls between cells. First-generation systems have no separate signaling channel to do this. Instead, they rely on a blank-and-burst technique that interrupts the speech channel for periods of about 100 ms, producing clicks in the speech path. They also tend to be expensive, restricting their use to business customers, and power-hungry, which tend to restrict them to permanent installation in a car.

### 5.3.2 Second-Generation Systems

Second-generation systems are characterized by digital transmission of speech and by more sophisticated signaling. The standards for these systems all make use of speech compression to reduce the data rate for speech to below the 64 Kbps that is the norm for inside digital telecommunications networks. The data rate is reduced to reduce the radio bandwidth required for operation, since radio bandwidth is a scarce resource.

Cellular systems use a greater level of speech compression than cordless systems, because less bandwidth is available to them, since their greater range of operation limits the reuse of bandwidth. In addition to using compression to reduce the data rate, they also use more sophisticated modulation techniques to increase the number of bits per second transmitted per hertz of bandwidth. Their more efficient speech compression requires a longer speech sample to be analyzed before transmission, producing a significantly greater coding delay than for cordless systems.

In the second-generation systems, there has been a convergence of cellular and cordless functionality. The direction for cellular systems was to reduce the cost and size of the cellular systems to be closer to cordless systems, and to use a dedicated out-of-band signaling channel in a similar way to a cordless system. Second-generation cordless systems have moved to support the transfer of ongoing calls between different base stations and to achieve the mobility of handsets, which is characteristic of cellular systems.

The need to transfer ongoing calls in a second-generation cordless system has been questioned, because there appears to be less need for this in an office environment than in a moving vehicle. Cellular systems are useful because they enable communications when the user is traveling. When a user is at home or in a business office, the amount of mobility required is much less. On the other hand, developments such as the increased use of roller skates by the users of cordless systems are now covered.

Different techniques are used to achieve duplex operation in second-generation standards. In cordless systems, *time-division duplex* (TDD) is used. This is another name for ping-pong or TCM in some digital copper and optical-fiber access systems. In cellular systems, *frequency-division duplex* (FDD) is used, with a different frequency band used for each direction of operation. (See Table 5.2.)

**Table 5.2**
Transmission Characteristics of Second-Generation Systems

| | *Cellular* | | *Cordless* | |
|---|---|---|---|---|
| *Characteristic* | *GSM* | *IS-54* | *CT2* | *DECT* |
| Frequency (MHz) | 890–915 + 935–960 | 824–849 + 869–894 | 864–868 | 1880–1900 |
| Duplex technique | Frequency | Frequency | Time | Time |
| Number of carriers | 2 times 125 | 2 times 800 | 40 | 10 |
| Carrier spacing (kHz) | 200 | 30 | 100 | 1728 |
| TDM channels per carrier | 8 simplex | 3 simplex | 1 duplex | 12 duplex |
| Voice data rate (Kbps) | 22.8 | 13 | 32 | 32 |
| Capacity (channels) | 1,000 | 2,400 | 40 | 120 |

The capacity of cellular systems to handle simultaneous calls is about an order of magnitude greater than that of cordless systems. The number of terminals that can be handled is even higher in comparison, because the larger the number of terminals handled, the lower the probability that a significant fraction will be active at once because of statistical averaging.

The framing structures used by second-generation systems differ from those normally used in telecommunications. Different forms of multiplexing into time slots, frames, and multiframes are used. Likewise, there are also differences in the layer 2 formats used for control and signaling.

### 5.3.3 Second-Generation Cellular

The speech compression used in the second-generation cellular standards is based on the analysis of blocks of samples that are 20 ms long. The delay produced by waiting for a complete sample block to be received before it is transmitted alone makes it necessary to use echo cancelers to avoid perceptible and annoying echoes in the speech paths.

The output of the speech coding of the standards contains two priorities of bits (see Figure 5.2). The higher priority bits are protected by adding additional error detection and forward error correction bits. The European GSM coder outputs a 13-Kbps stream, but error detection and forward correction increases this to 22.8 Kbps, which is actually transmitted. Likewise, the U.S. IS-54 coder's output at 7.95 Kbps is increased to 13 Kbps before transmission, and the Japanese standard coder's output is increased to 11.2 Kbps.

Although second-generation cellular systems have a separate control channel that is distinct from the speech channel, they still use blank-and-burst to usurp the speech channel during handoff between cells, because this allows the handoff to be completed more quickly. The clicks this produced in the first-generation systems have been eliminated in the second-generation systems by interpolation that estimates the speech samples that have been lost.

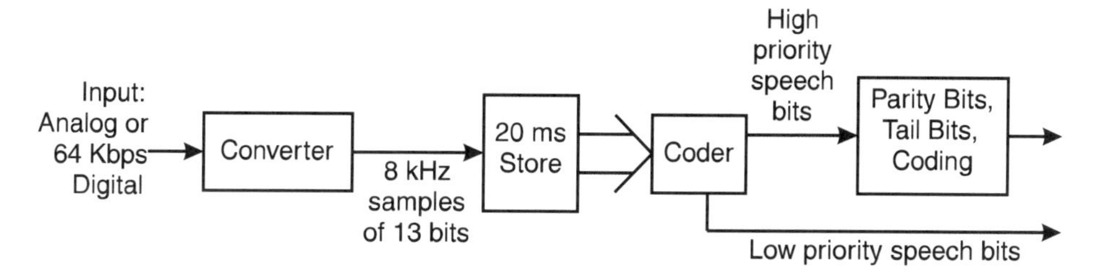


**Figure 5.2** Speech coding for second-generation cellular systems.

The capacity, in terms of active channels, of the European GSM systems is lower than that of the U.S. and Japanese standard systems, because the latter are forced to use smaller bandwidths per channel to be compatible with the corresponding first-generation systems. In Japan and the United States, the analog first-generation systems can be upgraded to digital operation on second-generation systems using the same radio spectrum channels.

#### *5.3.3.1 GSM Systems*

The pan-European GSM standard was intended to allow a single handset to be used for mobile telephony anywhere in Europe. It uses 125 carriers in both of its frequency bands, one band for each direction of operation. Carriers are spaced at 200-kHz intervals, and each carries both the speech and the signaling for eight terminals using TDMA, as is used in many optical-fiber systems. This allows a maximum of 1,000 active calls to be supported in every cell. Coordination between cells is based on CCITT Signaling System #7.

The TDMA is structured into multiframes of 120-ms duration, which consist of 26 frames (see Figure 5.3). The 13th and 26th frames are used for system control communications and the remaining 24 frames are used for speech traffic and associated signaling. Each of the 24 traffic frames contains eight time slots, one for each terminal.

Speech is sampled and analyzed in blocks 20 ms long, which are represented as 260 bits, consisting of 182 high-priority bits and 78 low-priority bits. Error detection and forward correction bits are added to the high-priority bits to give a total of 456 bits. Compressed speech from two blocks, 912 bits in total, are interleaved to spread out the errors due to bursts of noise and so prevent bursts from swamping the forward error correction. Two groups of 57 bits of interleaved speech are transmitted in each time slot, separated by bits carrying associated signaling. This processing allows the compressed and inter-

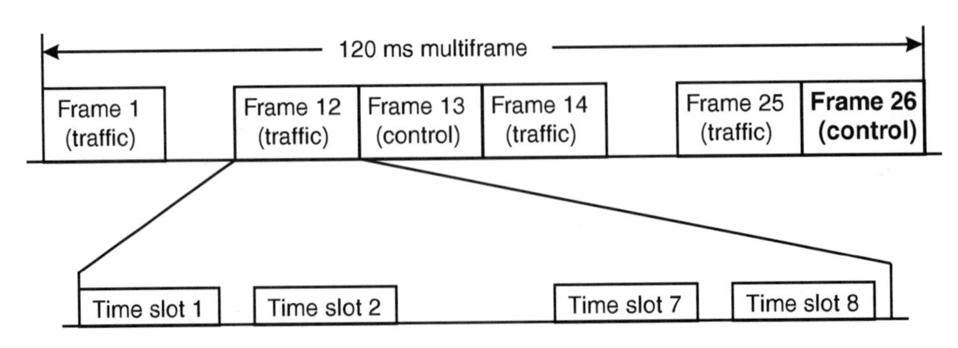


**Figure 5.3** Frame structure for GSM.

leaved speech bits from 40 ms of voice signals to be transmitted over eight traffic frames of the multiframe. Although the processing significantly improves the immunity of the speech to bursts of noise, it results in delays of 40 ms in each direction, making echo cancelers essential for acceptable operation.

GSM also contains a number of separate control and signaling channels, both for call setup and for system control. These communications channels are relatively slow, so GSM also allows speech over four frames, about 20 ms, to be interrupted when a fast communications channel is required. This interruption represents an acceptable degradation in the quality of service.

Because of the speed of its adoption, both within Europe and elsewhere, GSM is sometimes taken to stand for "Global System for Mobile communications."

#### *5.3.3.2 IS-54 Systems*

The Electronic Industry Association Interim Standard IS-54 is functionally equivalent to the GSM standard, but is designed to be consistent with the first-generation AMPS standard adopted in the United States and Canada. Like GSM, it has one band of frequencies in each direction, but it has significantly more carriers than GSM, about 800.

The IS-54 carriers are spaced at 30 kHz and correspond to the carriers used for AMPS. Like GSM, TDMA is used to support multiple terminals on each carrier, but IS-54 has only three terminals per carrier instead of GSM's eight. The overall capacity of IS-54 is greater than that of GSM, since it can support about 2,400 active calls per cell.

A relatively simple framing structure is used for IS-54. Frames are 40 ms long and contain six time slots, with two time slots per frame used for each customer. A different format within the time slots is used for each of the two directions of transmission.

Like GSM, speech is sampled and analyzed in blocks 20 ms long, but these are represented as 159 bits, consisting of 77 high-priority bits and 82 low-priority bits. The error detection and forward correction bits added to the high-priority bits give a total of 260 bits for transmission. Two blocks of 260 bits are interleaved to improve performance and transmitted on the two time slots in the frame which are allocated to the terminal. This produces a 40-ms delay in each direction, as for GSM, again making error cancelers essential.

The IS-54 control and signaling channels are separate, but are based on the AMPS protocols. However, there is the added capability to switch between digital and analog operation. Like GSM, speech traffic can be interrupted if faster communications is required. IS-54 uses a more efficient modulation technique than GSM to reduce the bandwidth required for the transmission of each bit.

#### 5.3.3.3 Japanese Systems

The Japanese standard is similar to IS-54, but has even less bandwidth available because it has to interwork with a first-generation approach using 25-kHz spacings. To achieve this, it uses increased voice compression and yet more efficient modulation.

### 5.3.4 Second-Generation Cordless

Second-generation cordless systems break the one-to-one relationship between handsets and base stations that characterize first-generation cordless systems The relevant standards have been designed to allow several handsets to operate from a single base station and to allow the same handset to be used at different base stations. The intention in breaking the one-to-one relationship was to allow wireless PABX operation in a business environment, with a number of handsets being usable over an extended area, covered by a number of base stations. There was also the intention to support public access base stations (Telepoint) at appropriate locations where anyone with a suitable handset could make a call, avoiding the need to wait for a public phone booth to become free.

Although second-generation systems were also targeted at the residential market, the ready availability of lower cost first-generation products has made the take-up low. Increasing sales in the business and Telepoint markets should result in lower prices and greater demand. The higher quality of cellular systems and the ability to support several terminals from a single base station, together with their ready availability at reasonable cost, has led to the development of a market that was originally unforeseen. This is their use inside a conventional telecommunications network to replace the last link from a distribution point to a customer's premises. This application is given momentum because it facilitates the liberalization of the telecommunications network by helping operators provide service without installing an extensive and costly copper infrastructure.

Second-generation systems avoid the problems of interference between terminals using the same channels by having automatic selection of interference-free channels. This also eliminates the margin required in designing a system with a fixed assignment of channels, because the reuse of channels is decided locally according to the actual local conditions. If a channel does start to suffer from interference, then the call can be handed over to a different channel between the terminal and the base station (intracell handoff). There is also the potential for a call to be handed over to a different base station (intercell handoff) that gives cordless telephony a similar ability to support mobility as cellular telephony, although the "cells" for cordless telephony are much smaller. The need for intercell handoff for cordless telephony has been questioned, because the practical applications of cordless and cellular systems differ.

Additional encryption and authentication procedures are required between terminals and base stations to ensure there is adequate security when different terminals have access to the same channels. It is also necessary for terminals to register with a local base station, because second-generation systems break the one-to-one relationship between terminals and base stations.

Duplex operation in second-generation cordless systems is achieved by the technique of TDD. This is another name for ping-pong or TCM in copper and optical-fiber systems. This approach for cordless systems differs from the FDD approach used for cellular systems.

Although speech compression is used for both cellular and cordless, CCITT standard 32-Kbps ADPCM is used for cordless systems. This has a much lower delay, 1 to 2 ms, than that used for cellular and so does not in itself require the use of echo cancelers. The delay introduced by the DECT frame structure does require echo cancellation to be used, but this is not the case for CT2 systems.

Although there are functional similarities between DECT and CT2, it is unlikely that DECT will replace CT2, because CT2 is well suited to its primary application of speech telephony. CT2 will not prevent the emergence of DECT systems, because DECT has considerably greater functional capabilities than CT2. In particular, DECT has the ability to support Basic rate ISDN operation, but it is not obvious that its functional advantages will ever enable it to gain a sufficiently high volume market to make it cost-competitive with CT2. If CT2 were to be extended to handle ISDN, it could easily make DECT obsolete.

#### 5.3.4.1 CT2 Systems

The CT2 standard supports good-quality voice and modem operation in a simple and effective way. It uses 40 carriers in a single frequency band with the same carrier being used in each direction at different time periods to support a maximum of 40 active calls at each base station. Carriers are spaced at 100-kHz intervals.

The channel is formatted into frames that are 2 ms long and that contain two time slots, one for each direction (see Figure 5.4). Time slots are separated by a guard time. Each time slot contains 64 bits of CCITT 32-Kbps ADPCM encoded speech, corresponding to 2 ms of encoded voice. The short frame length and short encoding delay make it unnecessary to use echo cancelers.

Each time slot also contains either 2 or 4 bits for control and signaling. Control messages consist of between one and six code words, each code word containing 48 information bits and a 16-bit *cyclic redundancy checksum* (CRC) for error detection. If a receiver detects an error in the CRC, it requests a retransmission of the word. There is no interruption of the speech channel by control or signaling information.

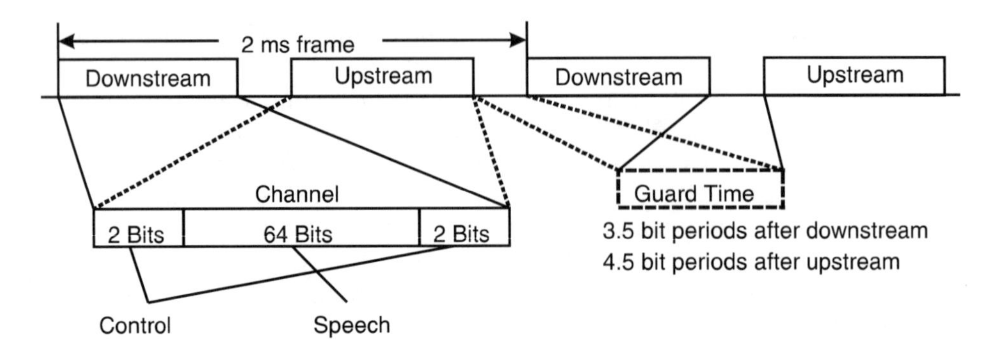


**Figure 5.4** Frame structure for CT2.

CT2 systems were widely implemented, especially in the United Kingdom. They were designed to meet well-focused requirements, supporting high-quality voice transmission in a simple, low-cost way. This and their ready availability led to them being tested in trials and accepted both within Europe and elsewhere.

### 5.3.4.2 DECT Systems

The goal of the DECT standard was to support a range of enhanced capabilities for cordless communications. In particular, it was intended to support basic rate ISDN operation. The result is a much more complex standard than CT2, and one which requires more sophisticated and expensive terminals.

DECT uses 10 carriers in a single frequency band, just below 2 GHz, which is about twice the frequency of operation of the other second-generation systems already described. Like CT2, it uses TDD to achieve duplex operation, so the same carrier is used in each direction at different time periods. Unlike CT2, each carrier supports 10 terminals using TDMA in a similar way to second-generation cellular systems (see Figure 5.5). A single DECT base station is able to handle 120 telephony calls simultaneously, but fewer calls can be handled if some of the calls require full basic rate ISDN transmission.

The carriers are spaced at 1,728-kHz intervals and are formatted into frames that are 10 ms long. Each frame contains 24 time slots, 12 in each direction, with the channels separated by guard times. Two time slots per frame, one for each direction, are allocated to each of the 12 PSTN terminals the channel supports. Fewer ISDN terminals can be supported when they have both B-channels active.

The bit periods are shorter for DECT than for other second-generation systems, because it operates with the highest data rate per channel. This can restrict its use, because its shorter bit periods are more likely to be corrupted by *intersymbol interference* (ISI) due to multiple transmission paths.

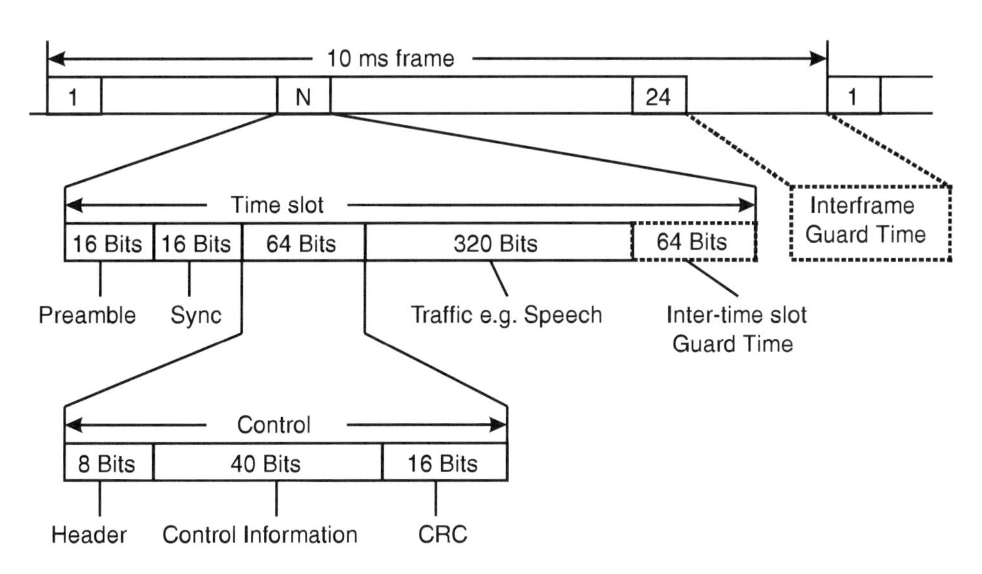


**Figure 5.5** Frame structure for DECT. Time slots 1 to 12: downstream. Time slots 13 to 24: upstream

Each time slot contains 320 bits of CCITT 32-Kbps ADPCM-encoded speech, corresponding to 10 ms of encoded voice. This is five times the amount transmitted in a CT2 frame. The frame assembly delay of 10 ms is greater than the speech encoding delay, and makes echo cancelers necessary to avoid degrading the quality of service. Each time slot also contains 64-bit control words that include 16 error detection bits, similar to the control words transmitted over several CT2 frames for CT2. There is also similar retransmission of control words in which errors are detected, but the higher layers of the protocol are more complex and its data rate is higher. There is again no interruption of the speech channel by control or signaling information.

The development of DECT systems has lagged behind that of CT2 systems, because the standard was agreed on later and because DECT systems have a longer development phase due to their greater complexity. The DECT approach clearly demonstrates the technical virtuosity of the experts who devised it. It is less clear whether the frame structure that forces the use of echo cancelers when it could have been avoided can accurately be described as sensible.

## 5.4 APPLICATIONS OF RADIO TECHNOLOGY IN ACCESS NETWORKS

There are three principal applications of radio technology in access networks. These applications are characterized by the portion of the complete link between the customer and the exchange carried by radio link. In certain systems, the entire link is carried by radio. In others, radio is used only for part of the link, either at the exchange end or at the customer end.

If radio is not used for the full link, then the link is typically divided at the distribution point. Radio can be used for the link between the exchange and the distribution point for remotely located communities, where it could be expensive to install a copper cable. Radio can also be used on the link between the distribution point and the customer for speed and ease of deployment, especially if there is competition between different operators.

The link from the exchange to the distribution point is typically between one and two orders of magnitude longer than the link from the distribution point to the customer. If radio is used for the full link, then the applications are similar to its use from exchange to the distribution point (i.e., rural or isolated customers). A full link can also be used on a shorter distance and at a higher rate to a business customer, often because this allows service to be provided very quickly.

Although this type of business application may be logically point-to-point to provide a single telecommunications channel, it is more appropriate to use a point-to-multipoint system, because it is more economical. Point-to-multipoint operation is realistic because there are more customers than distribution points and more distribution points than exchanges. The different applications are better differentiated by their distances of operation and end points than by their logical topologies.

If radio systems are used, then this should be transparent to the customers. It is desirable to keep the transmission delays low to avoid the additional complexity of adding echo cancelers and the annoyance produced by excessive delays, even when the echoes have been eliminated. This consideration constrains the length of the frames and the degree of speech compression. Long frames introduce delay because encoded speech has to be buffered between time slots in adjacent frames. Highly compressed speech introduces delays because long blocks of speech must be analyzed to give high levels of compression. High levels of compression of speech also interfere with modem operation, again reducing the desired transparency.

The two types of duplex operation, FDD and TDD, use similar amounts of total spectrum, but TDD uses a single band shared in time between both directions of transmitters, whereas FDD uses two bands. If spectrum is allocated on demand, to make better use of scarce bandwidth, then TDD may have an advantage in extreme cases because it may be easier to find one free band instead of two, but this is likely to be a marginal consideration.

FDD may have an advantage on long lines because the bit periods are longer, making it less susceptible to ISI from multiple paths. This argument is even stronger when used against TDMA, because TDD approximately halves the bit periods, whereas TDMA can reduce it by much more. FDD has the disadvantage that it requires a duplexing circuit to allow simultaneous transmission and reception on a single antenna.

### 5.4.1 Remote Multiplexers

A radio link can be used to carry telephony to a remote multiplexer serving a small, isolated community. In this application, a single radio transceiver provides service to a number of customers so that the cost overhead of the radio link is shared. This radio link typically provides the connection between a remote multiplexer and the telecommunications network (see Figure 5.6).

Customers are connected to their host multiplexer by normal copper pairs. The radio link in this configuration may be 10 km or more and only the final connection to the customer, perhaps a few hundred meters, is on copper. The radio transmission system may be point-to-point if only a single remote multiplexer is served, or it may be point-to-multipoint if a single system at the exchange serves a number of multiplexers. Repeaters may also be used on the radio link if necessary.

The use of radio for remote multiplexers is analogous to the use of fiber in FTTC or FTTCab applications.

### 5.4.2 Full Loop Systems

In full loop radio systems, the final copper drop to the customer has also been replaced by the radio link (see Figure 5.7). These are sometimes called *one-per-customer* (1pC) radio systems because, in contrast with a remote multiplexer systems, it is necessary to install single radio terminals for exclusive use by each customer. To avoid confusion, the term 1pC will not be used further here, since other systems can also have a single dedicated radio terminal per customer.

A full loop system can have an advantage over a radio link to a remote multiplexer, because it eliminates the cost and delay of installing the multi-

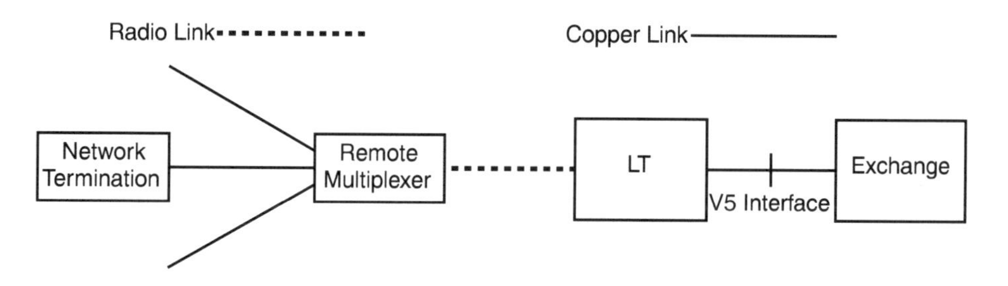


**Figure 5.6** Remote multiplexer application.

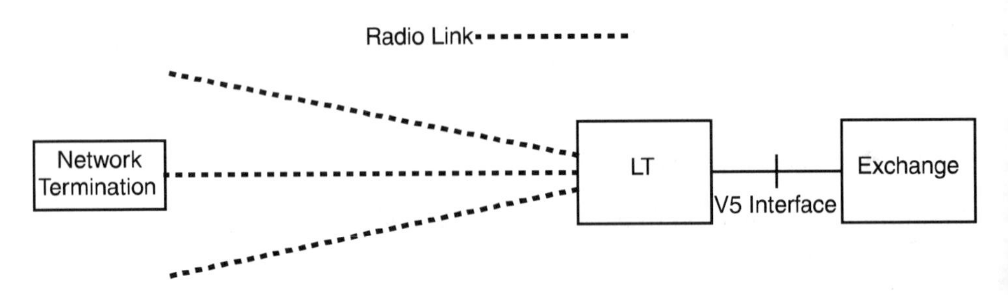


**Figure 5.7** Full loop application.

plexer and the associated copper pairs to the customers. It also eliminates the associated maintenance costs.

It is also effective to use a point-to-multipoint configuration for the radio system, because this eliminates the need for separate customer terminations at the exchange for each customer. A full loop system typically consists of one *radio line termination* (RLT) located at the host exchange and a number of *radio network units* (RNU) located at various customer premises. The distance between the RLT and an RNU may be up to 15 km to support operation in rural areas.

TDMA multiplexing should be avoided for full loop system customers, because the frame delay can result in the need for echo cancelers and the shorter bit periods are more prone to ISI from multiple paths at longer distances. This leaves FDMA and CDMA as potential options. CDMA operates by using a high-speed pseudorandom signal as a carrier, which is modulated by the data stream. A number of different pseudorandom signals can operate simultaneously in the same frequency band if they are chosen to be mutually orthogonal. CDMA has the property that it degrades smoothly as the number of sources increases, and this may be an advantage. It has the disadvantages that the number of orthogonal signals is limited and the complexity of a CDMA implementation is higher than an FDMA one.

Full loop systems are increasingly seen as cost-effective alternatives to providing service in remote, rural areas. They are of particular interest to operators who are obliged to provide service quickly in these areas, especially since there is less up-front investment required, since the RNUs need only be installed when service is required. This ensures that there is an immediate return on investment in the RNU.

A full loop radio system is analogous to an FTTH optical system.

### 5.4.3 Radio Drops

In terms of architectural structure and range of operation, a radio drop is the complementary application to the use of radio to provide a link to a remote multiplexer. Radio to a remote multiplexer replaces the first half of the loop between the exchange and a distribution point, whereas a radio drop replaces the second half between the customers and the distribution point (see Figure 5.8). Radio to a remote multiplexer is typically used in rural applications, whereas radio drops may be effective when used in urban applications. Radio to a remote multiplexer requires long-range operation, whereas a radio drop requires short-range operation.

Radio drops and full loop radio systems both require the installation of a dedicated radio terminal at each customer's location, and both eliminate the cost and delay of installing a final copper or fiber drop to the customer. The ease and speed of installation make them both suitable for providing service without the need for an expensive infrastructure, providing a fast return on investment.

Radio drops can also be used where there already is a copper infrastructure. They can support additional demands for service, such as a second line to an existing location when there is insufficient capacity remaining in the existing infrastructure. Radio drops can also be used as an overlay in areas of high traffic turnover (churn), even if the existing infrastructure has sufficient capacity, because they are more flexible than the fixed drops. They can also be useful if the existing infrastructure belongs to a different operator, because they provide a simple way of picking up customers from the other operator.

It is more important for bandwidth to be allocated on demand for radio drops, because in the urban environment there are potentially more demands for the scarce radio bandwidth. The range of operation in urban areas need only be a few hundred meters because of the high density of population. This range can be achieved with simple low-power systems, particularly if the radio terminals are fixed just below the eaves of the buildings, although this eliminates the potential advantage of limited mobility within a residence.

Further advantage can be taken of the concentration associated with the allocation of radio bandwidth on demand by carrying this concentration back into the network. The infrastructure to support radio drops can be kept lean if the concentrated and compressed voice traffic is transparently carried back to

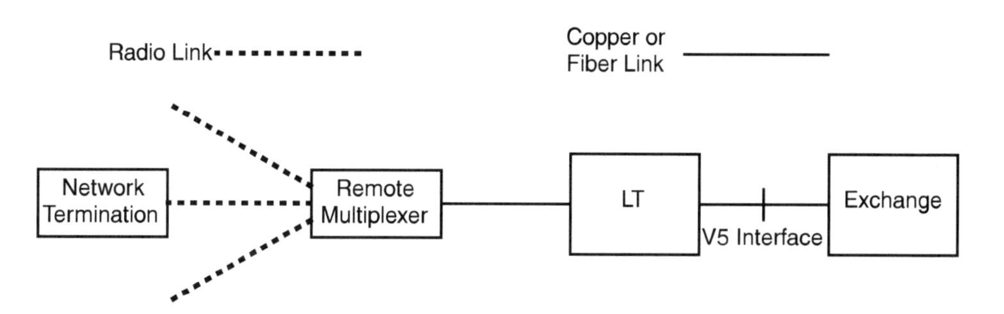


**Figure 5.8** Radio drop application.

the exchange. Keeping the infrastructure lean can be of particular advantage to new operators who are in competition with existing ones. The cost per customer of the interface to the exchange can be kept down if this can also handle concentration and compression.

ISDN basic rate technology can be used for the digital link to the exchange, because it is readily available and operates on almost all lines. Copper pairs will also support 1.544- or 2.048-Mbps links to the exchange either using traditional technology or HDSL. Optical transmission can also be used between the exchange and the radio base station at the distribution point, for instance, as a type of FTTC configuration. The basic rate approach may have an advantage because it is simplest and can allow the radio base station to be powered from the exchange, avoiding the need for battery backup.

It is natural to use point-to-multipoint operation for radio drops, because it simplifies the design of the radio base station. The costs of the implementations can also be reduced if advantage is taken of the high volumes associated with cordless radio systems, especially if the cordless technology can be used directly. Cordless technology can also be used indirectly at higher power on different frequencies to extend the range of the radio drop. The range of operation can also be extended by more sophisticated antennas than those appropriate to conventional cordless operation.

## 5.5 THE ADOPTION OF RADIO TECHNOLOGY IN THE ACCESS NETWORK

The GSM and IS-54 second-generation cellular systems are not ideal for access network applications because of their high levels of speech compression. Echo cancelers would be required because of their compression delays, and the compression algorithms restrict the data rate achievable with modems. There may be some scope for their expedient use where the degradation in the quality of service is acceptable in return for a fast provision of service.

The CT2 and DECT cordless systems do not suffer from these problems, because they use more modest speech compression, but DECT systems still require echo cancelers because of the delays introduced by their frame structures. Cordless systems are also more suitable than cellular for radio drop applications because the characteristic functionality of cellular systems to support mobile operation is not required. In particular, a radio drop system that is implemented with cordless technology can be kept simple and at low cost because it is not necessary to transfer ongoing calls between base stations. There is also some benefit in the radio terminal being able to automatically work with any base station, because this eliminates the issue of compatibility of terminals and base stations and encourages cost competition between suppliers.

A typical cordless application can be based on CT2 to provide basic telephony and be upgraded to DECT to support basic rate ISDN. A total of 30 customers can be supported by six CT2 channels, because this gives an acceptable quality of service with the automatic concentration produced by allocation of bandwidth on demand. Use of basic rate transmission allows the base station to be located up to 5 km from the exchange and still be powered from it.

Ideally, the interface to the exchange would use a V5.2 interface because it further reduces the cost by allowing the concentration on the radio system to be carried back into the exchange. A V5.2 interface does not support speech compression, but this produces less of a saving than the concentration that it does support. In theory, a V5.1 interface that was modified to allow concentration could be very effective, because it would be simpler to implement than a V5.2 interface. This is not an appropriate approach, because the amount of traffic it could support would make duplication of the interface link desirable, increasing the cost and complexity.

Radio terminals situated near the base station may need to have attenuators fitted to eliminate intermodulation interference. Those that are near the boundary served by an adjacent base station may need to have a directional antenna to avoid interference. Without directional antennas, there is a reduction in the number of terminals that can be served, because the same frequency band cannot be used for both base stations simultaneously.

The use of standard cordless systems for radio drops is restricted by the need to coexist with other users of cordless systems. The domestic use of cordless systems in the area covered by a radio drop is especially limiting, because it is difficult to maximize the use of the bandwidth through synchronized operation. If too many cordless systems are used in one area, then interference may result in failures to establish calls, or even loss of calls in progress. These types of failures may need to be monitored to ensure that the quality of service is maintained.

## 5.6 SUMMARY

Access networks can take advantage of the recent developments in cellular and cordless radio. Radio links may be used for the full link between a customer and an exchange, or for a partial link if used in conjunction with copper or fiber transmission systems. They may also be used for expediency or as a permanent solution.

Cellular radio systems support mobile communications by handing over the connection between the base stations of adjacent cells as the customer moves between them. Initially, cordless systems had a dedicated base station for each customer located at the customer's premises; however, the differences in functionality between cordless and cellular systems has decreased as the technology has moved from analog first-generation systems to digital second-generation systems. Despite the increasing similarity, cellular systems have a greater range and capacity and use narrower bandwidths per voice channel.

There are no globally accepted standards for first-generation systems, and even within Europe there is no agreement on the adoption of the CT1 cordless specification, because some countries considered it to be too complex. The need for simplicity was again discounted when the second-generation DECT cordless specification was produced in Europe, but at least the simpler European CT2 specification for PSTN was also agreed to. Fortunately, there has been widespread adoption of the European GSM system for second-generation cellular radio, both within and outside of Europe, although different approaches compatible with the bandwidths of the first-generation systems have been agreed to for the United States and Japan.

If radio is used to replace the full loop or for a link to a remote multiplexer, then cellular systems are more appropriate because of the longer range necessary. A radio link to a remote multiplexer can be useful for providing service to a small isolated community, but it is preferable to use radio for the full link, since this avoids the need to power and maintain the remote multiplexer. However, it may be better to deploy second-generation cellular systems as an expedient solution for rapid provision because of their lower quality of service. This is due to the high level of speech compression used to keep the channel bandwidth down, which introduces delays making echo cancelers necessary and limiting the data rates for modems.

Second-generation cordless systems, especially CT2, could be used very effectively on the last drop between a remote multiplexer and a customer, because this avoids the cost of the cable infrastructure. DECT is less effective because it is more complex and because echo cancelers are still necessary, due to the DECT frame structure, although the speech coding does not require them. The use of cordless technology for radio drops is especially important in a competitive environment because it makes it easier for an operator to provide service in an area that is dominated by another operator.

## Selected Bibliography

Arnbak, J. C., "The European (R)evolution of Wireless Digital Networks," *IEEE Communications Magazine*, Vol. 31, No. 9, September 1993, pp. 74–82.

Begley, M., E. A. J. Kempton, and N. Sharp, "One per Customer Digital Radio System," *Fourth European Conf. on Radio Relay Systems*, October 1993, pp. 90–96.

Benvenuto, N., G. Bertocci, W. R. Daumer, and D. K. Sparrell, "The 32-kb/s ADPCM Coding Standard," *AT&T Technical J.*, Vol. 65, September/October 1986, pp. 12–22.

Donovan, P., "Introduction of Radio Technology Into the Local Loop," *SUPERCOMM/ICC '92*, Vol. 4, June 1992, pp. 1828–1832.

Duet, D., J.-F. Kiang, and D. R. Wolter, "An Assessment of Alternative Wireless Access Technologies for PCS Applications," *IEEE J. Selected Areas in Communications*, Vol. 11, No. 6, August 1993, pp. 861–869.

Goodman, D. J., "Second Generation Wireless Information Networks," *IEEE Trans. Vehicular Technology*, Vol. 40, No. 2, May 1991, pp. 366–374.

Goodman, D. J., "Trends in Cellular and Cordless Communications," *IEEE Communications Magazine*, Vol. 29, No. 6, June 1991, pp. 31–40.

Howett, F., "DECT Beyond CT2," *IEE Review*, Vol. 38, No. 7-8, July 1992, pp. 263–267.

Huish, P. W., and S. A. Mohamed, "Radio Systems in BT's Access Network," *SUPERCOMM/ICC '92*, Vol. 4, June 1992, pp. 1823–1827.

Jones, D. L., "Fixed Wireless Access: A Cost Effective Solution for Local Loop Service in Underserved Areas," *1992 IEEE International Conf. on Selected Topics in Wireless Communications*, June 1992, pp. 240–244.

Lin, S. H., and R. S. Wolff, "Basic Exchange Radio—From Concept to Reality," *SUPERCOMM/ICC '90*, Vol. 1, April 1990, pp. 52–58.

Mannisto, H., "Wireless Local Loop—A New Access Instrument for Changing Europe?" *IEE Colloquium on Customer Access—The Last 1.6 km*–4, June 1993, p. 8/1.

Merrett, R. P., R. Warburton, and A. M. Rolls, "Wireless Access Based on CT-2 Radio System," *Fourth European Conf. on Radio Relay Systems*, October 1993, pp. 79–83.

Owen, F., and C. Geoffray, "The DECT Standard in Local Loop Access Applications," *Fourth European Conf. on Radio Relay Systems*, October 1993, pp. 103–107.

Pradhan, B. D., "Wireless Rural Communications for Developing Countries," *1992 IEEE International Conf. on Selected Topics in Wireless Communications*, June 1992, pp. 254–256.

Searle, R., "Personal Communications Services—Issues for WARC 92 in Europe," *GLOBECOM '91 Conf. Record*, Vol. 3, December 1991, pp. 1787–1790.

Sollenberger, N. R., and A. Afrashteh, "An Efficient TDMA Radio Link and Architecture for Wireless Local Access," *ICUPC'92 Proc.*, September 1992, pp. 10.02/1–5.

Taylor, J. T., "PCS in the U.S. and Europe," *IEEE Communications Magazine*, Vol. 30, No. 6, June 1992, pp. 48–50.

Tuttlebee, W. H. W., "Cordless Personal Communications," *IEEE Communications Magazine*, Vol. 30, No. 12, December 1992, pp. 42–53.

# 6 The Powering of Access Networks

"Power corrupts."

—Anonymous

For access networks, it may well be that the things that are most corrupted by power are the grand visions that people have had about the future. In one case at least, the problem of powering has been blatantly dismissed as a mere detail. In the real world, before the advent of reliable cold fusion, antimatter containment, or the tapping of the quantum vacuum, it may be a less trivial problem.

The discussion of powering given here is at a higher level than that of the transmission technologies, since the purpose of this chapter is to identify the problems and their possible solutions.

## 6.1 THE TRADITIONAL BALANCE

In the traditional approach to PSTN, the telephone handsets are powered from the exchanges to which they are connected. The exchange supplies dc power across the same pair of copper wires that carries the voice signal. There is no need for the power to be supplied by the customer unless a special terminal, such as a fax or an answering machine, is used. The PSTN service is also made independent of the mains electricity supply by having a backup power source, typically an independent generator, in the exchange building.

It is important for PSTN to continue to operate during a mains failure because voice communication is needed during emergencies when mains power is not available. The PSTN service was designed to be independent of local mains power, because local mains power was not universally available when the PSTN service was developed. Likewise, the catchment area of an exchange is limited in part by the ability of the exchange to power the handset, and this influences the thickness of the copper wire used, since thinner wire is sufficient for lines that are close to the exchange. It is preferable to use thin copper wires because they cost less than thicker wires and because they take up less space in the underground ducts. This puts an upper limit on the gauge of the copper wires. The more expensive thicker wires are used on the longer lines, where the total resistance must be kept low if sufficient power is to be available for the handset.

In the traditional approach to PSTN, a balance exists between the distance from the exchange to the handset and the ability of the exchange to power the handset, especially in the event of a mains failure. The catchment area of an exchange must at least be large enough to contain enough customers to support the use of a standby generator. This places a lower limit on the gauge of the copper wires used, since power must be transmitted to the boundaries of this catchment area. The constraints on voice transmission do limit the size of the catchment area for traditional telephony services, because the ability to transmit voice signals increases with the ability to transmit power. This is because both the loss of power and the attenuation of the voice signals are proportional to the total resistance of the wire, ensuring that a pair of wires that can carry power can also carry a voice band signal.

From this it should be clear that traditional PSTN has evolved to a natural limit where the distance from the exchange to the handset matches the ability to transmit power to the handset, and the ability to carry information is ensured by the ability to carry power. This is not a natural limit for new technologies, since the catchment areas are determined by the distances over which they can transmit information and not by considerations of powering. The balance between power and catchment area no longer exists.

## 6.2 PROBLEMS FOR NEW TECHNOLOGIES

One of the first challenges to the balance within the traditional approach to telephony came from attempts to take advantage of the underutilization of the information-carrying potential of existing copper pairs. Attempts to exploit this potential are most obvious now in the development of HDSL and ADSL systems, but the initial challenge came from pair-gain technology.

It is expensive to install a new copper cable when all of the spare pairs on the existing cable have been used. It is even more expensive if this has to be done in reaction to an unforeseen demand for service if the installation of new cables was planned at a later date. It is also unnecessary to install new cables, because the existing cables have unused information-carrying capacity, and it is only the number of pairs that have been exhausted.

Pair-gain systems and multiplexing systems, both concentrating and nonconcentrating, have been devised to take advantage of the unused information-carrying capacity of copper pairs. Pair-gain systems need to power the electronics at the far end of the copper pair. Fortunately, it has been possible to take advantage of the bursty nature of the demand for power by using a rechargeable battery to buffer the power consumption. Although the peak power demand may exceed the capacity of the copper pair, the average power demand is likely to be within its capacity, since the line is more often inactive than active. A rechargeable battery can be used to supplement the power feed from the exchange when the line is active (see Figure 6.1).

In practice, the deployment of a rechargeable battery in this way creates problems, since the heavy use of the line by a business customer can completely discharge the battery. To some extent this can be compensated for if the installation is tailored so that the most heavily used termination is powered conventionally with the result that a battery is only needed on less heavily used terminations, but this is not reliable, because a customer is free to alter the use of these terminations. In addition, the limited life of a rechargeable battery, even in the benign environment of a customer's premises, leads to significant maintenance costs, especially if there is no warning of the death of a battery and an unscheduled maintenance visit has to be quickly arranged.

Unfortunately, it is not possible to use a battery to buffer the power consumption for more sophisticated and power-hungry pair-gain systems and for higher capacity HDSL and ADSL systems. This is because the average power for these systems typically exceeds that which can be delivered over their copper pairs. The problem is not as great as for radio and fiber-optic systems because the transmission of sufficient power to the remote terminations by radio or fiber is not feasible. Even so, copper systems that use a battery to buffer the power consumption of remote terminations may not have a significant advantage over optical-fiber and radio systems because of the cost of maintaining and replacing the remote battery.

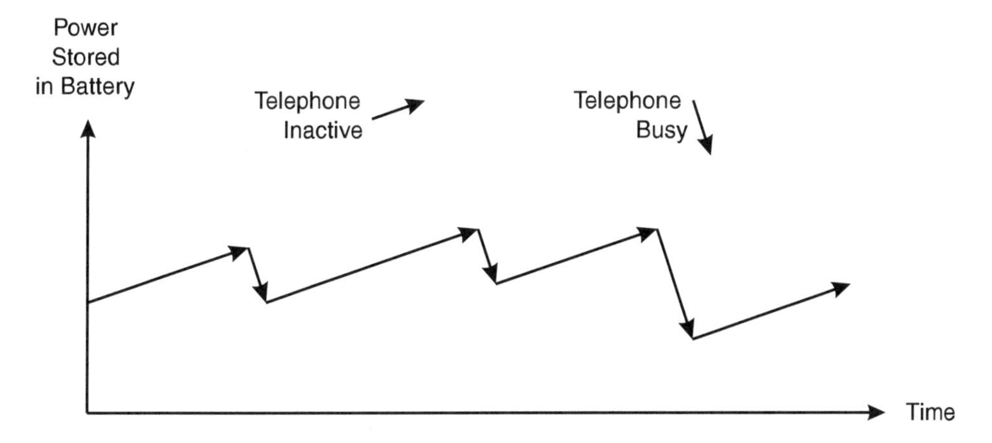


**Figure 6.1** Use of a rechargeable battery to buffer the power consumption.

At first sight, radio appears to be the worst medium for powering, because no significant amount of electrical power can be conveyed across a radio link. In contrast, optical fibers can transmit a useful amount of energy, and it has been suggested that this could be sufficient to power a special low-power handset. It has also been suggested that hybrid fiber/copper cables could be used to convey both information and power. Unfortunately, there are some fallacies in these concepts. Although a certain amount of optical power can be conveyed down an optical fiber, the low conversion efficiency into electrical power makes the necessary transmitted power so high that it could be hazardous during maintenance operations. In addition, a buffer battery would probably also be required so that the optical power can be reduced from the peak needed to nearer the average, reintroducing the problem of battery maintenance and heavy usage. Furthermore, if the cost advantages of passive splitting are to be gained, then the transmitted optical power would need to be increased beyond that required for point-to-point operation, because of the reduction of received power caused by the splitters. The concept of a hybrid cable is also misguided because copper systems are not even able to supply sufficient power to match their own information-carrying capacity, far less that of an associated optical fiber.

There is one genuine application of optical fiber to powering, but this application is ironic. The upper limit on the data rate of optical-fiber systems can be raised if optical amplifiers are used to boost the optical signal. These amplifiers can be powered very effectively over optical fiber, because their primary requirement is for optical, not electrical, power. This is ironic because the most appropriate use of optical power is one that increases the information-carrying capacity of optical-fiber systems, and so aggravates the problem of powering the remote terminations, because these need more power, since the capacity of the system has increased.

## 6.3 BATTERY BACKUP

If the local power in a telephone exchange fails, then emergency power is often provided by a backup generator. These generators are readily available and can be maintained in a benign environment, often with maintenance staff present. Generators can also produce more power for their size than rechargeable batteries.

It is often not practical to use backup generators at the far ends of access networks, even if these generators are small, because of the difficulties of ventilation and maintenance. Ventilation may be a problem because the far end may be contained in an enclosed space, such as an underground box. Maintenance is also a problem because it is necessary to send a maintenance technician to the remote sites that have been affected by a power failure to refill the fuel tanks after the generators have been used.

Rechargeable batteries are the obvious solution to the problem of backup power at the far ends of an access network. These batteries can be kept fully charged during normal operation and used to power each far end of an access network if the primary power supply fails. Nonrechargeable batteries could be used as an alternative to rechargeable batteries for one-shot operation. Like backup generators, nonrechargeable batteries require a maintenance technician to be sent out to replace the batteries after they have been used. Nonrechargeable batteries also have a limited shelf life and so would need to be replaced when they became too old, even if they had not been used. Rechargeable batteries are preferable because they have lower maintenance costs, since they do not need to be replaced after they are used.

Unfortunately, there are a number of problems that make the use of rechargeable batteries far from attractive. Perhaps the most physically obvious problem is the size of the batteries. The factors that determine the size of the batteries are the power requirements of the far ends and the duration for which power must be supplied, together with the power density, which determines how effectively the batteries store power. For PSTN operation, the power requirements are dominated by the power needed by the handset, which is fixed, and by the power dissipated by the line circuit. Although low-power line circuits are feasible, they require an apparently backwards step in technology, because they make use of old-fashioned passive transformers rather than chic integrated circuits, which dissipate more power. The duration for which power is supplied is determined by the statistics of local power failures and customer usage, which can be heavier during a power failure. Several hours of backup power at normal usage rates is typically a minimum. The rule of thumb is that the size of the rechargeable batteries is comparable to the size of the rest of the electronics.

There are also environmental constraints on the use of rechargeable batteries, and these constraints are likely to become more pronounced over time due to the increasing awareness of environmental issues. Sealed lead-acid batteries, such are commonly used in cars, are the most common type of rechargeable batteries. The lead these batteries contain make them potentially environmentally hazardous, and there is a requirement that old batteries be collected and disposed of safely. This is not a problem for the lead-acid batteries used in cars, because car batteries return to a garage that acts as a collection point when the cars are serviced. Aging lead-acid batteries in the far ends of access networks do not make their own way to a collection point, and so maintenance expense is incurred.

There is a similar problem with rechargeable nickel-cadmium batteries, because cadmium is also an environmentally hazardous material. Nickel-cadmium batteries are more difficult to use for power backup, because, unlike lead-acid batteries, there is no simple way to determine the power remaining in them, since their output voltage is almost independent of the remaining power stored. It is possible to compensate for this by using more sophisticated charging circuitry, but this adds to the complexity of the design. Nickel-cadmium batteries are better suited to active use rather than being kept charged to provide backup. Some designs use dual batteries that are alternately charged and discharged to ensure that a charged battery is always available, but this adds to the expense and the size.

Not all rechargeable batteries contain environmentally hazardous materials. Rechargeable lithium batteries have been developed, but they are not a feasible alternative because they are not readily commercially available. The rechargeable nickel-hydride batteries being marketed as an environmentally friendly alternative to nickel-cadmium batteries are a more attractive alternative.

However, the greatest problem with rechargeable batteries is the cost of maintaining them. Rechargeable batteries have been used for several years to power the far ends of pair-gain systems. Like more sophisticated access networks, pair-gain systems also suffer from the technological imbalance between the ability to carry information and the ability to carry power. There have been persistent problems with the batteries used for pair-gain systems, and the lesson from this has been to eliminate the batteries if possible to reduce the maintenance costs, even if this requires more expensive low-power electronics.

Rechargeable batteries also have a limited life span. Those used in cars are unlikely to last more than five years. Preliminary estimates suggest that batteries used in the far ends of access networks may have life spans of nearer two years. This means that if it is not feasible to eliminate them, it will be necessary to send out a maintenance technician to between 20% and 50% of all far ends every year to replace their large rechargeable batteries. Worse, the initial lifetime of batteries in some places has been found to be even less that two years, but fortunately the lifetime increases after an initial period, because the local inhabitants discover that these batteries do not fit into their cars. However, if batteries are widely deployed, it is quite possible that battery lifetimes will fall again, since human resourcefulness is likely to ensure that an appropriate battery adapter is designed.

The operational cost necessary to maintain rechargeable batteries could easily outweigh the cost of the active electronics at the far ends, because the operational cost will be incurred every year over the entire lifetime of the equipment.

## 6.4 REMOTE FEEDING

The primary source of power for the far ends of an access network is almost always the mains electricity supply, because alternative sources of power are appropriate only in special circumstances. The source of the primary power supply can be either collocated with the equipment being powered or remote from it. The first case is called *local powering* and the second case is called *remote feeding* (see Figure 6.2). It is important to clarify this distinction further, because otherwise there is the question of how far away the primary supply can be before it is classified as remote. Powering is classified as remote if the transmission of power to the equipment is a significant issue.

In these terms, the traditional powering of telephone handsets can be called *remote feeding*, because it can have a significant impact on the size of the catchment area of a local exchange. The fact that there is often a large physical distance between the exchange and the handsets is not the principle consideration, because it is the loss of power over distance and not the distance in its own right that is important. The invention of a room temperature superconductor could make all distances local in terms of powering.

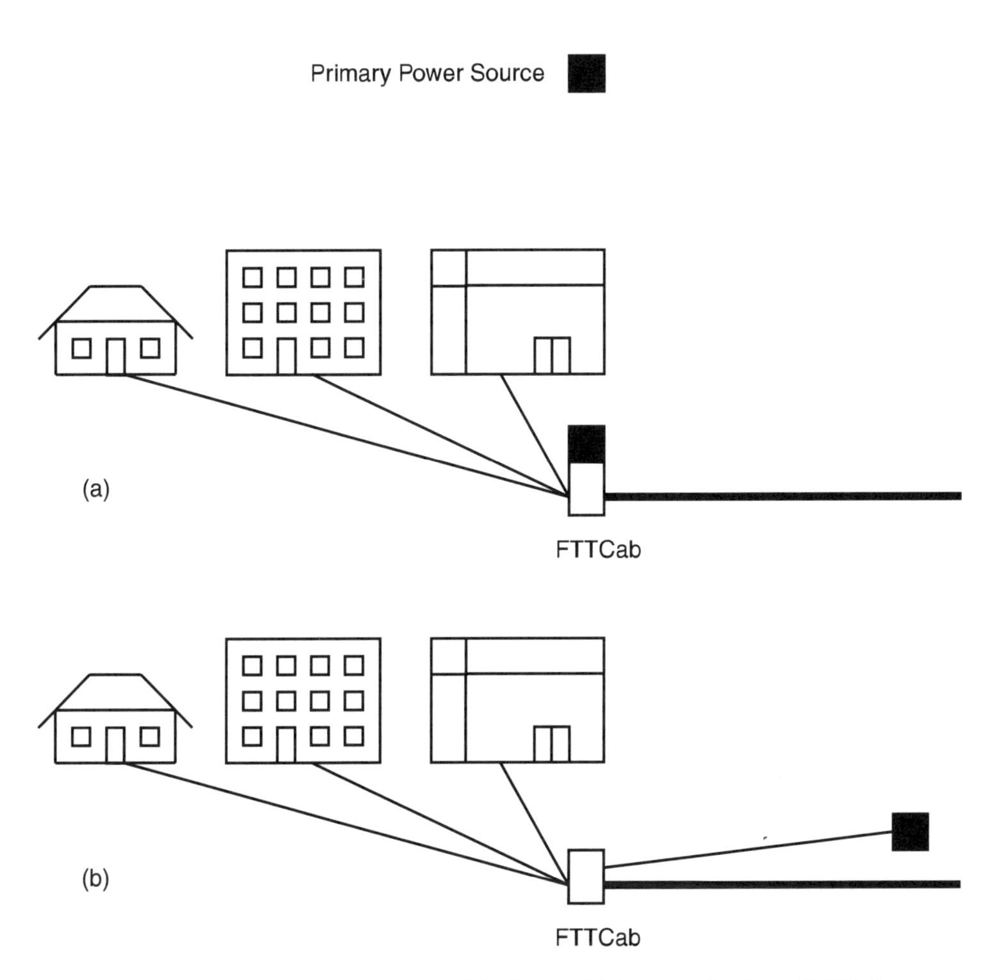


**Figure 6.2** Example of (a) local powering of FTTCab and (b) remote feeding of FTTCab.

Remote feeding can take two forms (see Figure 6.3). The more obvious form is the fan-out configuration, where a single primary power source feeds a number of remote sites. The other option, the fan-in approach, has the direction of the power feed inverted with a number of primary sources feeding a single remote node. The fan-out approach feeds power to nodes of the network or to customer equipment. The fan-in approach has been considered because it was seen to have the potential of allowing an access network work with power supplied by the customers.

The equipment sited at the host exchange is unlikely to need remote feeding, because it can be powered from the same primary source as the exchange itself. For simple access networks, this equipment will be the headend of the access network. If the headend is sited distantly from the exchange, then both it and the remote end of the transmission system that feeds it will require power. These could be remotely powered from the exchange using a fan-out configuration, and this approach would also be applicable to a two-stage access network with an active intermediate node.

Power is also required at the far ends of an access network and for the customer's equipment. For radio and optical-fiber systems, the far ends are

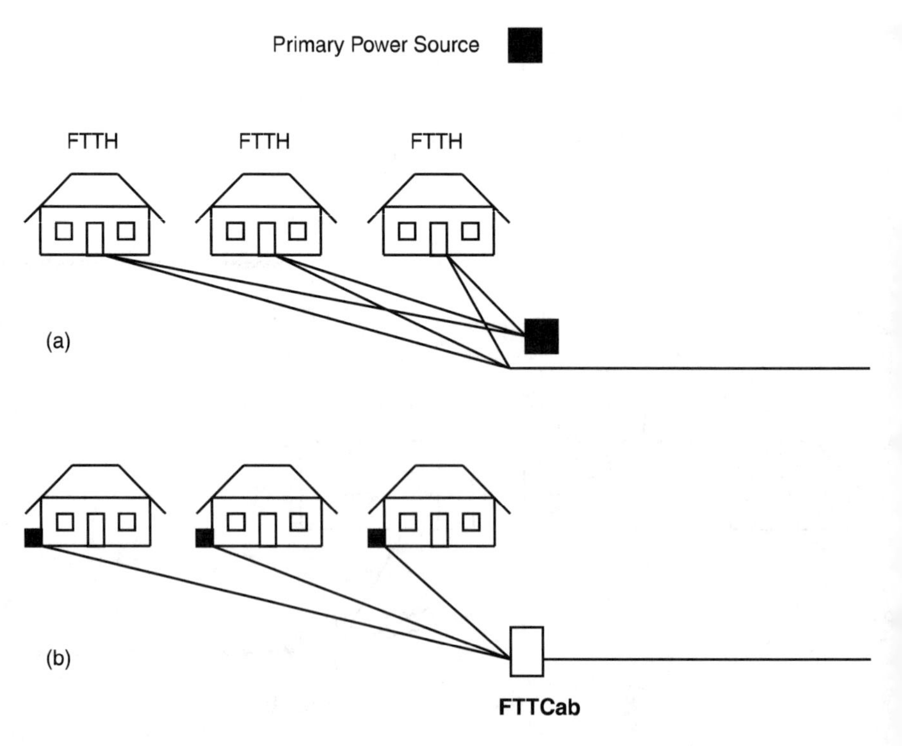


**Figure 6.3** Remote feeding: (a) fan-out powering of FTTH and (b) fan-in powering of FTTCab.

likely to require primary power supplies that are independent of those that support the headends, because it is difficult to transport sufficient power over these physical media. If the far ends of radio and fiber systems are in the customer's premises, then remote feeding is unlikely, because it would require the installation of specialized remote networks for powering that would duplicate the existing mains distribution networks.

Remote feeding at the far ends of radio and optical systems is more appropriate for business, curb, cabinet, or apartment systems, because these can have a common far end that serves a number of terminals such as telephone handsets. These types of far ends may either use a fan-out configuration to remotely power the terminals from the common far end or a fan-in configuration to power the common far end from the terminals. Remote feeding of radio systems is not as appropriate as remote feeding of fiber systems because it undermines the avoidance of a fixed infrastructure that is a key advantage of radio systems. Remote feeding of the far ends at business and apartment locations is also less appropriate, since businesses are often prepared to provide local power for communications and both can provide a benign environment for local powering. The fan-in configuration for remote feeding is especially unsuited to business and apartment locations in comparison to local powering, since both depend on the presence of mains power in the building, but the fan-in approach requires many more mains units, resulting in much higher equipment and maintenance costs.

The fan-in configuration is better suited to curb configurations because of the difficulties of providing local power to a large number of curb locations. The advantage of the fan-in configuration is that the responsibility for powering can be delegated to the customers while ensuring that the failure of one customer to supply power does not bring the whole system down. There is a corresponding disadvantage that the burden of powering may not be spread evenly among the customers, and the cost of maintenance of the battery backup might not be eliminated.

The ability to remotely power far ends and distantly sited headends using a fan-out configuration is governed principally by the application of Ohm's law. The loss of power during transmission is determined by the resistance of the conductors used and by the current that flows in them. Special large-diameter cabling could be used to remotely feed the nodes of optical and radio systems, but this type of cabling is not appropriate to copper systems because it is preferable to use existing copper cables for copper systems. Special power cabling is also not desirable for radio and optical-fiber systems because it increase the installation costs. The alternative, especially for copper systems, is to increase the operating voltage of the power transmission. High voltage is used for mains power distribution because it reduces the current flowing and hence the power lost due to dissipation by the resistance of the copper wire. However, there are a number of factors that limit how far the voltage can be increased.

Safety may not be the greatest problem for high-voltage operation, because high voltage in its own right is not dangerous. Static build-up can often produce voltages high enough to cause electrical breakdown of air, and yet only cause minor annoyance to humans. If there are sufficient precautions taken to limit the current, then a high-voltage feed could be made adequately safely. A greater problem to the use of high voltage, especially on existing copper, may be the reduction in the reliability of the conventional insulation. This can be a problem because the insulation tends to break down after prolonged operation at high voltage. The use of both positive and negative voltages to increase the differential feed while reducing the voltage difference to earth suffers from a different problem of aging, that of induced corrosion. Negative voltages were chosen for traditional systems because these tend to inhibit corrosion, whereas metallic corrosion is encouraged if positive voltages are used. Differential, square-wave ac may be the optimal technical solution, with the square waves only rounded sufficiently to reduce the generated interference, since this avoids the problems of dc feed and minimizes the peak voltage.

It should be noted that the location of the primary power source does not determine the location of the battery used for backup, although batteries may need to be larger if they are not sited at the active nodes. However, batteries that are used as power buffers will need to be sited at the active nodes to reduce the transmitted power.

## 6.5 LOCAL POWERING

Pure remote feeding is unusual, because in addition to the remotely powered equipment, there is often equipment that is collocated with the primary power supply. For example, there may be local mains powering of the far ends of an access network, which in turn acts as a remote source of power for customer equipment.

If a local mains supply is available, then this may be preferable to remote feeding, because the advantage of remote feeding may be compromised, since although batteries may be needed to back up the local power supply, they may be needed anyway to buffer the power consumption if remote feeding is used. In theory, remote feeding provides independence from a local mains failure, but this assumes that the remote source is itself immune to a local mains failure. If the remote source is protected from local mains failure by battery backup, then larger batteries are required than if they were sited locally because of the power loss in transmission. In addition, if local batteries are needed anyway as a power buffer, it may be more attractive to increase the size of them and use local powering, because the batteries will have to be maintained in any event.

Remote feeding may beat local powering if it produces a saving in the costs of battery maintenance that outweighs the cost of the larger batteries required to offset power loss in transmission. Fan-in remote feeding of a far end shared by a number of customers may also produce savings in the cost of battery maintenance, since power backup can be centralized at the far end or delegated to the customers.

The greatest challenge for powering, when the location is inaccessible for either mains power or remote powering, may need to be solved by alternative sources of local power. Locations for which no local mains power is available can be remotely fed, possibly with batteries for power buffering. When this is not feasible, possibly due to the absence of a copper connection, then alternative forms of local powering are the only options. Solar power, wind power, and water power have been used as alternatives to mains electricity in awkward locations. Solar and wind power used in conjunction has been proven to be an effective combination, despite the high equipment costs. Water power has the advantage that it is more reliable and less likely to need a buffer battery, but it is typically more expensive because of the installation costs of a water-powered turbine generator. Flywheels are also being considered as an alternative to batteries in remote locations.

It has been suggested that local powering could be made independent of a local mains failure by using a remote feed as a backup. This is superficially attractive because it appears to eliminate the greatest drawback to local powering, that of local mains failure, without the need for a local battery backup. Unfortunately, there is a fallacy here in that if remote powering can operate successfully during an emergency, then it could also power the equipment for normal operation.

## 6.6 ALTERNATIVES TO BATTERY BACKUP

Traditionally, it has been convenient and sensible to ensure that telecommunications continue to operate during a local mains power failure. It has been convenient because power was supplied remotely from an exchange that could have its own generator. It is also sensible because it assists in dealing with emergencies, which are more likely to occur due to the power failure. This traditional approach needs to be reviewed because of the difficulties of providing backup power.

The first step in this review is to question the need for backup power for telecommunications. There is unlikely to be a need to provide backup power for many new telecommunications services, because these new services have no tradition to preserve and because end customers already accept the loss of services if there is a local power failure. At the very least, end customers are probably willing to accept a reduction in telecommunications services during a failure.

Many end customers are probably even willing to accept a loss of PSTN service during a local power failure, because customers are often surprised to learn that PSTN service is not normally affected. This acceptance by end customers might not imply that PSTN service can be allowed to fail, because national telecommunications regulators may insist continued operation is required to help deal with emergencies. Even if national regulators do not object, it may still be unwise to allow PSTN to fail because of the "dead granny" factor; that is, there could be strong adverse publicity if a vulnerable member of society died as a consequence of the loss of PSTN service during a mains failure.

If certain telecommunications services need to be immune to a local power failure, then alternative approaches could be adopted. Power could still be provided locally, but the responsibility for this could be devolved to the customer. Large business customer may already have their own backup generators, which could be used to power the required telecommunications equipment. Smaller business customers and private residential customers might need to provide battery power, and this could be supplemented by including a battery alarm on the customer premises to warn of immanent battery failure. Similar battery alarms are already used on smoke detectors, where the consequences of failure could also be serious.

Another alternative would be for the end customer to use a hand-crank to generate sufficient power to make an outgoing call in an emergency. A similar arrangement was used in early telephones to generate sufficient power to ring a bell. With suitable publicity, this could even be portrayed as attractively quaint, and it is quite clearly environmentally friendly. Obviously, this approach might not be suitable for a business office, where something resembling an executive exercise bicycle might be more appropriate.

A more serious alternative could be to supply a small radio handset for use in emergencies. This could simplify the devolution of the responsibility for backup power to the customer, because this would reduce to ensuring that there were good batteries in the handset. It has the further advantage of supporting full or partially mobile communications in other emergencies, and could cost less to provide than the cost of the maintenance of rechargeable batteries.

A further alternative is for the telecommunications operator to make provision of telecommunications services during a local power failure into an optional extra within the service agreement. This approach would allow the customer to chose whether or not to have a backed-up service, and to pay the associated costs if required. This could be especially attractive as an option for new services if it made the inconvenience of a local power failure more tolerable.

## 6.7 SUMMARY

The traditional approach of powering PSTN handsets is well suited to the service requirements and to the technology. The copper pair that carries the traffic is also capable of supplying power to the small number of handsets at its far end. The catchment area of a traditional exchange, which is limited in part by the power dissipated by the resistance of its copper pairs, is large enough for an emergency power generator to be used in the exchange building, and so avoids the problems of battery backup.

The new technologies being deployed in access networks need new approaches to powering of their far ends, because it is much easier to transmit information than it is to transmit power. The power required at the far ends is also greater for the new technologies, because the far ends contain active electronics. In contrast, the traditional approach had passive far ends and only needed to supply power to the telephone handsets.

Advanced copper technologies are best suited to remotely powering the far ends because power can be transmitted on the copper pairs. Unfortunately, the power required at the far ends is typically greater than the limit imposed by the resistance of the copper. This limit can be overcome by operating at a higher voltage, but this is constrained by the dangers and increased aging introduced by the higher voltages. Similar considerations apply to the use of additional copper to remotely power the far ends of radio and optical-fiber systems. Ironically, optical fiber can be used to carry optical power to pump optical amplifiers, but this aggravates the problems of remotely powering the active far ends, because it further increases the capacity of new systems to transmit information.

The alternative to remotely feeding the far ends of an access network is to power them locally. This is normally achieved through local mains powering, but alternatives are needed if this is not available, and solar and wind power used in combination has been effective in extreme cases. Local powering is vulnerable to a local power failure, and protection against this typically requires remotely located rechargeable batteries.

There are a number of problems with rechargeable batteries. The size of the batteries would double the size of the far ends. The most common types contain lead or cadmium, which are environmentally hazardous materials and so have legal requirements imposed on their disposal. Worst of all, perhaps, is the associated maintenance cost due to their limited life spans, which can outweigh the cost of the batteries.

The simplest solution to the loss of local power may be to simply accept that this will mean the loss of services other than PSTN. It may even be possible to gain acceptance for the loss of PSTN service also, because most people are surprised to learn that telephones still operate during a mains failure. If it is essential to maintain an emergency PSTN service during a mains failure, then alternative solutions may need to be sought. These alternatives could include hand-cranking, as used on some of the earliest telephones, or alarms to alert users to failing batteries, as for smoke alarms. It may even be effective to supply a radio handset for emergency use, since this could cost less than maintaining the rechargeable batteries.

## Selected Bibliography

Dixon, J. S., and P. Best, "Renewable Energy Supplies for Radio Systems for Servicing Rural Customers," *Colloquium Record: IEE Colloquium on Radio Systems for Rural Communications*, January 1993, pp. 6/1–4.

Fisher, S., "Powering Active Nodes in Active Loops," *ICC'91 Conf. Record*, Vol. 2, June 1991, pp. 929–35.

Kuhn, D., E. Lo, and T. Robbins, "Powering Issues in an Optical Fibre Customer Access Network," *Conf. Record: INTELEC '91*, November 1991, pp. 51–58.

Lo, E. W., "Cost Analysis of Powering an Optical Customer Access Network," *Conf. Record: INTELEC '92*, October 1992, pp. 96–103.

Mistry, K., "Powering Fiber-in-the-Loop Systems," *IEEE LTS*, Vol. 3, No. 4, November 1992, pp. 36–44.

Salloum, H. R., and K. J. Mosian, "Fiber in the Loop Systems Powering: Bellcore Requirements," *Conf. Record: INTELEC '92*, October 1992, pp. 117–123.

# 7 ATM in the Access Network

"Technology which has brought meaning to the lives of many technicians."

—adapted from Ed Bluestone

The deployment of access networks using copper pair and radio technology differs from the deployment of those using optical fiber, since copper and radio are already established media. More advanced copper systems will be deployed by traditional operators in particular applications because of the advantage of using an existing copper infrastructure, and these may support ATM, particularly for new services such as VoD, maybe easing the eventual transition to optical fiber. Radio systems will continue to be deployed and will be especially useful, since they allow service to be provided without the need for a cabled infrastructure, and they may also support ATM. However, it is optical-fiber systems that will provide the greatest opportunity for the deployment of ATM in the access network, although it may be in conjunction with coaxial cable, especially in the United States.

The discussion of ATM in the access network is made more difficult because a large number of acronyms are used. Readers may wish to refer to the list of acronyms and abbreviations in the back of this book if these acronyms become confusing.

## 7.1 AN OVERVIEW OF ATM

ATM is an approach to transmission, switching, and multiplexing aimed at creating a communications fabric that is common across a number of different supporting technologies. The basic unit of ATM communications is a constant-length packet, known as an *ATM cell*, which consists of 5 bytes of control information, known as the *ATM header*, followed by a 48-byte payload (see Figure 7.1).

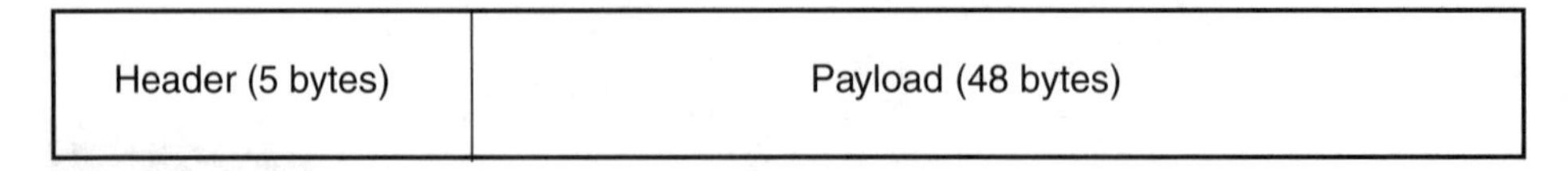


**Figure 7.1** ATM cell structure.

### 7.1.1 Virtual Paths and Virtual Channels

The first 4 bytes of the header of the ATM cells contain a *virtual path identifier* (VPI), a *virtual channel identifier* (VCI), and certain other information fields (see Figure 7.2). The VPI acts as the high-order address field and the VCI as the low-order address field for the ATM payload. Different payloads can be identified by the 3-bit *payload type* (PT) field in the header. The fifth byte of the header is a checksum that allows corruptions of the first 4 bytes to be detected.

ATM cells at a point in the network that have the same VPI belong to the same *virtual path link* (VPL). A *virtual path* (VP) switch may route ATM cells according to the ports and the VPI labels of the cells at its ATM interfaces and may change the VPI labels in the process. All the cells with the same VPI value from the incoming VPL must be switched to the same outgoing VPL by a VP

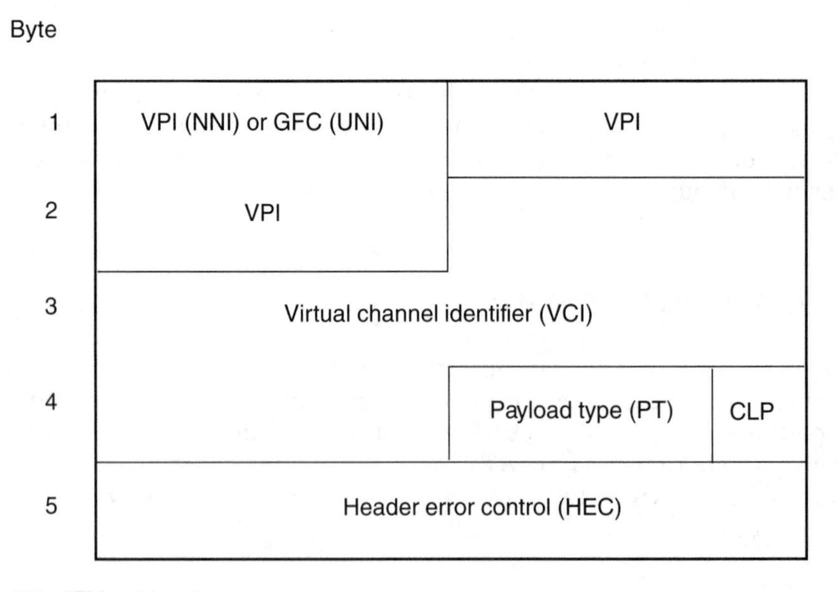


**Figure 7.2** ATM cell header.

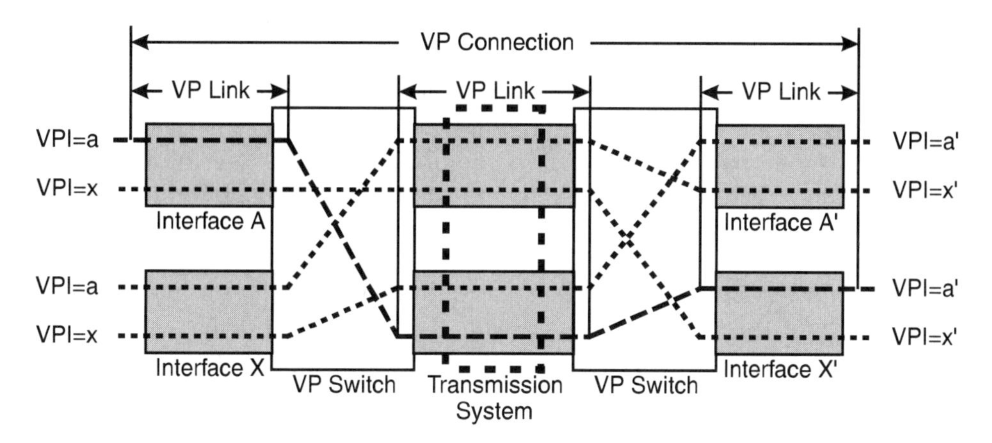


**Figure 7.3** Virtual path links, switches, and connections.

switch. A VP switch cannot change the VCIs of the ATM cells or route cells according to the value of their VCIs, because these functions are performed at the *virtual channel* (VC) level, not at the VP level. A sequence of connected VPLs is known as a *virtual path connection* (VPC), and ATM traffic input into one end of a VPC emerges at the other end, but normally with a different VPI label.

ATM cells within the same VPL that have the same VCI belong to the same *virtual channel link* (VCL) that is served by that VPL (see Figure 7.4). Only VC switches can route ATM cells between VCLs according to the VCI labels of the cells, and the VCIs can be changed in the process. A *virtual channel connection* (VCC) consists of a sequence of connected VCLs, and ATM traffic input into one end of a VCC emerges at the other end, but typically with both the VPI and VCI labels changed. A VP or VC switch may have its routing table changed either on demand from a user, through call control signaling, or by intervention of the network operator by management control, in which case it is more precise to call the switch a *cross-connect*.

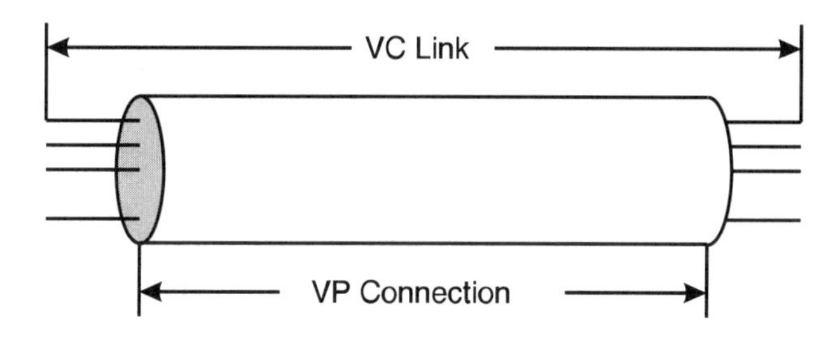


**Figure 7.4** VCLs served by a VPC.

### 7.1.2 Service Types and Adaptation

Services make use of ATM cells through adaptation of the service onto ATM cells. The service layer is the client of the *ATM adaptation layer* (AAL), which is in turn served by the ATM *virtual channel* (VC) layer. Several adaptation protocols have been defined. AAL1 is the adaptation protocol for *constant bit rate* (CBR) services and AAL2 is the adaptation protocol for *variable bit rate* (VBR) services. The combined AAL3/4 protocol is used for the *switched multi-megabit data service* (SMDS), which was defined principally for linking *local-area networks* (LAN) and *metropolitan-area networks* (MAN) at 1,544 or 2,048 Kbps. The AAL5 protocol is suitable for packet data applications, including signaling.

Additional service types for *available bit rate* (ABR) and *frame relay* (FR) services have also been defined. The ABR service type is intended for low-priority services that make use of spare capacity. The FR and SMDS service types are more in the nature of specifically identified applications, which have been distinguished from the other generic types. CBR services typically require guaranteed data rates, transmission delays, and jitter performance.

### 7.1.3 The Functional Architecture of ATM

The functional architecture distinguishes between the ATM layers and the physical transmission layers that serve them. This is important both for the management of the systems and for actual implementations. If implementations integrate the functions from these different layers, then the ability to adapt an implementation to operate with a different physical medium may be jeopardized. If no distinction is manifest in the functional architecture, then the management software that typically mirrors the functional architecture may be more expensive to implement because it is less modular.

In the conventional ATM functional architecture, the ATM layers, consisting of the VC layer and the VP layer that serves it, are served in turn by a transmission convergence layer (see Figure 7.5). The transmission convergence layer is in turn supported by the physical transmission layers, which may include the section and path layers of an SDH transmission system. This conventional architecture has to be extended if a multipoint transmission system is used in an access network.

In conventional point-to-point ATM systems, a single physical transmission system supports a single remote end, which supports a single user port. For multipoint operation in an access network, a single physical transmission system may support a number of remote units, which in turn support a number of ports. This added sophistication may require the addition of one or more architectural layers between the transmission convergence layer and the physi-

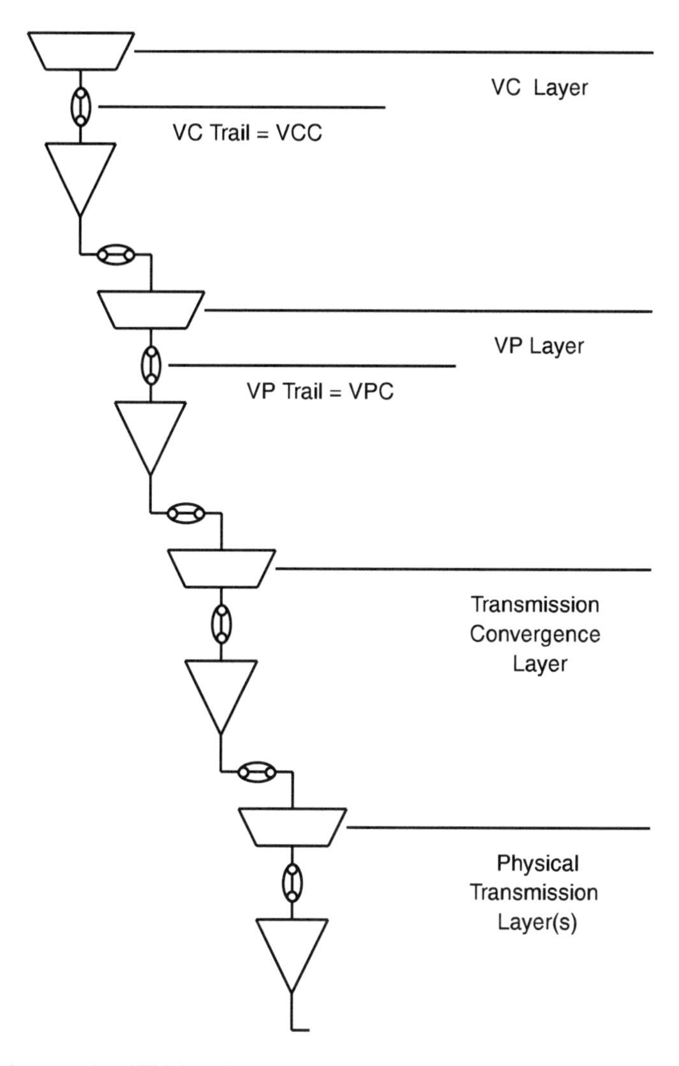


**Figure 7.5** Point-to-point ATM functional architecture.

cal transmission layer, since the transmission convergence layer must be immediately below the VP layer, as it is responsible for the ATM cell delineation.

## 7.2 ATM ON AN OPTICAL ACCESS NETWORK

In addition to the modifications to the functional architecture required for ATM to be used with multipoint transmission systems, there are further issues that are specific to the use of ATM on optical access networks.

### 7.2.1 Architectural Topology

Star architectures have already been introduced in the access network to provide fiber links to large business customers. It is likely that these will continue to be used for ATM in preference to shared fiber topologies, since they are already in place. When new fiber is used for small business customers and for residential customers, it is likely to be shared fiber with a passive multistar PON topology, since this reduces the cost of creating the new fiber infrastructure.

Fiber to the building, business, or home is preferable to fiber to the cabinet or curb for evolution to ATM operation on PONs, because these avoid the electronic bottleneck that constrains future broadband operation. ATM operation can, however, still be performed at reduced data rates, for instance, on existing copper pair drops. Although FTTC has been proposed for initial deployment of narrowband systems, this could result in increased costs when upgrading to ATM at more conventional rates.

### 7.2.2 ATM PON Transmission

The multiplexing of upstream ATM traffic from the remote ends of an *ATM PON* (APON) is simplified if it is done on a per cell basis rather than a per bit basis, since the synchronization of the remote ends to achieve bit interleaving is difficult at the rates required for broadband traffic. In addition, it is natural to use the intrinsic granularity of ATM cells as the granularity of APON transmission. A guard time of a few bit periods between the cells from different remote ends does not introduce much inefficiency, because it is small in comparison to the cell period. In addition to the guard time, a few bits of preamble can be used to allow the receiver to synchronize with the start of the incoming cell and so ensure that bits within the cell are sampled at the correct time.

In the downstream direction, it is possible to simplify the transmission because a single ATM OLT at the headend broadcasts to all of the ATM ONUs at the remote ends. Transmission in the upstream direction is more difficult because it is necessary to prevent collisions between the ONUs transmitting to the OLT. If the duplexing technique allows for continuous transmission in the downstream direction, then there is no need for guard times and preambles because each of the remote ONUs can continuously synchronize with the downstream flow. If the downstream transmission is not continuous, for instance, if TCM (ping-pong) is used for duplexing, then the synchronization of the remote ends to the downstream flow can still be simplified because each remote end can maintain synchronization lock by using information sent to other remote ends. However, downstream transmission that is not continuous is less efficient.

### 7.2.3 APON Cell Addresses

It is natural to use a cell-based PON transmission system for ATM operation, because the cells used for APON transmission can be easily mapped onto the ATM cells. Adding a label to the APON cells is also natural, because it allows traffic and control cells to be differentiated and assigned to particular ONUs (see Figure 7.6). This APON label is not strictly necessary, but if it is not used, then certain values of the addresses in the ATM cell headers may need to be allocated for control communication with the ONUs. By using an additional APON label, the ATM traffic can be handled more transparently, because there is no need to reserve special ATM header addresses.

If no APON label is used, then only ATM cell addresses are present on the APON and so at least one ATM cell address is needed for control communication between the OLT and the ONUs. If only one ATM cell address is used for OLT/ONU control communication, then some mechanism is needed to share this between all ONUs, for instance, the OLT could poll each of the ONUs in turn. Through control communication with the ONUs, it is possible for ATM cells on the APON to be assigned on demand to particular VPs and VCs at the user ports. This creates a more flexible structure than if a fixed mapping between user ports and ATM cells is used. The communication between the OLT and the ONUs is also needed for ONU-specific monitoring and maintenance functions.

It would be possible for APON cells to consist of an ATM cell plus an APON label that identifies both the ONU and the port on that ONU with which the ATM cell is associated. This approach has the advantage of placing no constraints on the values of the ATM cell headers at the user ports. Unfortunately, this would require the header to contain several bytes, which is an inefficient use of bandwidth, since most of the header address space at the user port is unlikely to be used. It is also unnecessary if each ONU is aware of the mapping between the combined ATM cell header and APON label and its own user ports and associated ATM addresses.

### 7.2.4 Service Multiplexing

There is an advantage to using different VPs on an ATM access network for different services, since this allows the signaling and traffic for a particular service to be grouped together and distinguished from those of other services.

**Figure 7.6** Labeling of ATM cells for APON transmission.

Different VCs within the VP may then be used for the various signaling and traffic channels particular to the service on that VP. This approach also allows different services to be routed between customers and service nodes by ATM access network acting as a VP cross-connect without it needing to analyze call-control information.

It may be necessary for the ATM access network to change the VPIs, since certain combinations of VPI and VCI are reserved for specific local functions, such as management, and cannot be routed across the access network. It may also be necessary for the VP addresses to be used to identify their user ports so that customers can be differentiated at the ATM switch or service node. These VP connections across the access network are likely to be semipermanent, because they provide links to the service nodes, but many of the VC connections are likely to be on-demand, since they may be established as specific service capabilities are required.

## 7.3 PROTOCOLS FOR MULTIPOINT ATM OPERATION

For the sake of clarity, the discussion of the multipoint protocols here has been restricted to APON transmission systems. Although similar considerations apply to other multipoint transmission systems, such as multipoint radio, other systems face different specific problems that would have to be addressed in detail if a complete account is to be given. APON systems are described because they are likely to be the most widely deployed.

### 7.3.1 Ranging

A ranging process is necessary to allow the remote ONUs to compensate for the different propagation delays between them and the OLT at the headend. Ranging is required if TDMA is used in the upstream direction, but may not be necessary if other techniques such as FDM or CDM are used for the multiplexing of transmissions from the ONUs. However, techniques that do not require ranging often require more expensive transmitters and receivers, since more sophisticated modulation is required.

Full ranging need only occur during installation or following a major fault, since the propagation delay on a PON does not vary greatly over time. In addition to full ranging, the ranging may need to be fine-tuned more regularly to compensate for slight variations in propagation delay, for example, due to annual temperature variations.

If the remote ALUs transmit long bursts, then the repeated fine-tuning process may not be necessary, because it is not necessary to synchronize the ALUs to within a fraction of a bit as for some narrowband systems. The common receiver at the OLT can synchronize with the start bits at the beginning of the transmission burst from an OLT and fine-tuning can be eliminated if the variation in propagation delay is less than the guard time between bursts. Slight variations can be monitored by the synchronization with the start bits of the burst, and if the variation becomes too large, then either a fine-tuning correction or full ranging can be initiated.

Often ranging is performed in two steps, because a single-step approach can take a long time due to the resolution required and the large variation in the propagation delays. The situation is aggravated if a large number of ALUs need to be ranged, because it must be possible to range the entire system in a reasonably short time. In the first step of ranging, a low-resolution estimate of the propagation delay is obtained and communicated to the ALU being ranged, which uses this as a correction. In the second step of ranging, a high-resolution measurement is made of the difference between the estimated and the actual delay. If only the low-resolution estimate of the propagation delay was made, then the guard time between the transmissions of different ALUs would have to be much greater to allow for the poor accuracy of the estimated propagation delays. If synchronization with the start bits of the bursts is not used as a fine-tuning correction, then the repeated fine-tuning can be performed by periodic high-resolution measurements of the difference between the estimated and the actual delays.

Potential ranging techniques are constrained, because the ranging of one ONU must not interfere with the normal operation of other ONUs. Typically, two techniques are used for this, either separately or in combination. In every case it is necessary to resolve the conflict that occurs if more than one ALU tries to range at the same time. In the first technique, a remote ONU wishing to range sends a continuous low-level pseudorandom signal that acts as low-level noise on the transmissions from other ONUs. The OLT can detect this low-level signal because the signals from the other ONUs average out over time. A pseudorandom signal is used because it has properties similar to noise and because it gives a sharp response when the receiver searches for the delayed signal. This approach is suitable so long as it is possible to prevent the simultaneous transmission of pseudorandom signals from a number of ONUs wishing to range at the same time from swamping the other operational ONUs.

The alternative approach is to dedicate a time interval for ranging in the upstream transmission frame. An ONU that wishes to range can transmit during this interval. In this case, there may still be a collision if a number of ONUs wish to range at the same time, but this collision does not interfere with the normal operation of ranged ONUs, since it is confined to a special time interval. The disadvantage of this approach is that a large time interval must be allocated because of the wide variation in propagation delays between ONUs, and this interval will normally be unused.

A hybrid approach uses a pseudorandom signal for low-resolution ranging controlled by the OLT and prevents a collision of ranging requests by polling the ONUs. This approach is good for low-resolution ranging because the detection time for the signal is less if the required resolution is low. High-resolution ranging can then be performed by transferring the transmission from the OLT being ranged to a time interval reserved for this purpose. This time interval can be much smaller if two-step ranging is used, because after the first step it is not necessary to handle a large variation in delays.

### 7.3.2 Leveling

The received signal levels from the remote ONUs are different due to variations in their transmitters and in the attenuation due to their different propagation lengths. This variation in signal levels causes a degradation in performance, because the thresholds that differentiate between different transmitted symbols vary from one ONU to another and because the receiver in the OLT has to operate with an extended dynamic range. This degradation is reduced if the different transmitters in the remote ONUs are coordinated so that the received signal levels at the OLT are similar.

The process, known as *leveling*, which coordinates the signal levels transmitted by the remote ONUs, has some similarities with ranging. It differs from ranging in that there is no need for the remote ONUs to transmit a special signal during the process, because the OLT is always aware of the received signal level. The OLT controls the leveling process by providing feedback to each ONU to request an increase or decrease in the transmitted signal level. Monitoring of this level can provide an early indication of a future failure of the transmitter in an OLT, because this will often be heralded by repeated requests to increase the signal level.

### 7.3.3 Media Access Control and Capacity Allocation for APONs

The multiplexing of the upstream transmissions from the remote ONUs is more difficult than the multiplexing of the downstream transmission because of the need to coordinate the remote ONUs to avoid collisions on the fiber. The OLT at the headend can transmit a continuous stream of cells the remote ONUs can monitor, synchronize with, and select the appropriate cells from. In the upstream direction, the transmission from all the remote ONUs must be controlled so that multiplexing slots are only used by one ONU at any time.

Related to the control of the multiplexing of the upstream transmission is the allocation of capacity to the ONUs. Capacity need not be allocated symmetrically, because certain applications, such as VoD, require more capacity in the downstream direction than in the upstream direction. Like multiplexing, altering the capacity in the downstream direction is easier than in the upstream direction, since the capacity is determined by the number of APON cells the OLT chooses to send with the appropriate destination address. Downstream APON cells can be explicitly addressed, for instance, using a combination of

APON label and ATM cell header to identify the destination ONU, the port on that ONU, and the VP and VC at that port. Downstream cells can also be implicitly addressed by using a separate communications channel to inform the ONU of the mapping between the APON cells and the destinations.

In practice, a combination of explicit and implicit addressing may be used. For example, the ONUs may be explicitly addressed by the APON label while the mapping between ATM cell header and the ports and port VPs and VCs is determined according to demand by communication between the ONU and the OLT. In this approach, the APON label indicates whether the information in the APON cell is destined for a user port or for the ONU itself. Information destined for the ONU itself supports the communication between ONUs and the OLT.

Once the explicit and implicit addressing is established, the OLT can transmit more traffic to a specific downstream destination by sending more cells with the appropriate APON label and ATM cell header. Transmitting more traffic from a specific ONU is more difficult, because the upstream transmission is shared between ONUs. Typically, an ONU must request the OLT to grant it more upstream capacity to avoid collisions with other ONUs.

Capacity allocation in the upstream direction can be controlled by the OLT informing each ONU of the upstream APON cell slots allocated to it. Each ONU always has at least one APON cell slot allocated to it, since it must be able to communicate with the OLT. The ONU can use its allocated capacity to request additional APON cell slots. If the ONU has more capacity than it needs, it can cease to use some of its APON cell slots and use its communications with the OLT to indicate that the slots are free.

The allocation and deallocation of capacity is similar to the resource management function at the higher ATM layers. This capacity allocation function must be present if the upstream transmission is to be used efficiently, since otherwise capacity must be allocated to ONUs regardless of their need for it. The communications between ONUs and the OLT may use explicit addressing or implicit addressing, which is established through a metasignaling channel, possibly on initialization of the ONUs. Metasignaling should not be used to establish the capacity allocation channel when a demand for additional capacity occurs, because this demand needs to be handled quickly. Likewise, while there is sufficient spare capacity on the upstream transmission, each ONU should maintain a buffer of unused capacity so that it can meet any sudden demands from its user ports without needing to wait for additional capacity to be allocated.

## 7.4 GENERAL PROBLEMS AND MISCONCEPTIONS

Despite the large amount of work that has already been done on ATM, there are still a number of problems and mistaken notions that may act as a barrier to the deployment of ATM in the access network.

### 7.4.1 Signaling for Narrowband Services

There is a widespread fallacy that the signaling that has been specified for broadband communications, Q.2931, will support existing narrowband services. This fallacy is based on wishful thinking, which is easy, rather than on logical analysis, which requires mental effort. To many it is "obvious" that broadband signaling must be able to handle existing narrowband services because it is has evolved from the signaling for them, and that anyone who suggests otherwise cannot have a proper understanding of broadband signaling. Unfortunately, this faith is misplaced, and the truth can be found by considering the V5 signaling for narrowband services.

First of all, it is not possible to map existing PSTN services from different countries into the same ISDN call control protocol, because existing PSTN services have different, and in some cases incompatible, call models. It was necessary to create the V5 PSTN protocol to solve this problem, and this is not part of Q.2931. Admittedly, it is possible to fit PSTN services into Q.2931 signaling similar to the way that all of Procrustes' guests could be made to fit into his bed, but this involves chopping off the parts that are too big.

The second difficulty arises from the activation and deactivation of ISDN basic rate interfaces. It was necessary to define the activation and deactivation messages in the V5 control protocol because they are not covered by Q.931 messages. The same is also true for Q.2931 messages.

The problem of supporting narrowband services with broadband signaling is often pronounced solved by stating that all that is required is a broadband-ISDN terminal adapter that maps the narrowband services into the broadband network. Even if this were not technically impossible for Q.2931 signaling as indicated above, it is not reasonable to assume that customers will be prepared to pay for a terminal adapter to provide the same services they can obtain at present without one.

### 7.4.2 Transport of Narrowband Services

There are further difficulties in transporting narrowband services over ATM, because it takes 6 ms to fill an ATM cell with 64-Kbps traffic. This cell assembly delay causes problems for PSTN, because it creates a round-trip delay, which can lead to a perceived echo. Although the echo can be suppressed, this is not ideal, because the suppression leads to additional expense. The situation appears to become worse if the ATM cells are expected to handle compressed speech, because this creates even longer delays, but paradoxically this can become easier, because speech compression produces delays of its own that can require the addition of echo cancelers in any case.

Alternatives to adding echo cancelers are to multiplex traffic from a number of ports onto a single ATM cell and to only partially fill the cells.

Partially filling the cells is not desirable because it is an inefficient use of transmission bandwidth. Multiplexing the traffic from a number of PSTN ports is more efficient, but the cells cannot then be routed like other ATM cells, because different parts of the cell have different final destinations. This multiplexing approach is better suited to a hybrid narrowband/ATM network, because the ATM cells containing PSTN traffic can then be routed to the narrowband part of the network, where they can be broken down and routed as narrowband traffic.

### 7.4.3 Addressing Ranges at the Network Interface

The ATM standards are not well suited to the network interface between an ATM access network and its host ATM switch, because they potentially limit the interface to supporting 16 user ports. The reason for this is that the standards allocate a very large address space to the user interfaces, but only allow 4 bits more of address space at the network interface. It has been suggested that the origin of this address allocation was that it was based on the architectures used for computing systems and that the architectures common in telecommunications networks were not considered in any detail.

In practice, this restriction on the addresses at the network interface need not be a problem because it is unlikely that each user will really need all of the 16 million separate channels permitted by the ATM UNI specifications. If each user is restricted to a mere 4,096 channels, then the network interface can support up to 64 thousand user ports. This does require additional functionality in an ATM access network, because it must act as a VC cross-connect, as well as a VP cross-connect, but this may be required in any case, because the same problem can occur in the APON transmission system if this uses pure ATM cell-based transmission. The mapping between VPIs and VCIs on the user interface and those on the network interface must somehow also preserve the special addresses at both interfaces reserved for particular functions.

### 7.4.4 The ITU-T Reference Model

The ITU-T reference model for broadband-ISDN, which is based on that for narrowband-ISDN, is not practical for small-business or residential customers because of its implementation costs. The reference model prevents the cost-effective support of existing services because of the need for a terminal adapter. In addition, the presence of both an NT1 and an NT2 is expensive, because two pieces of equipment linked with a sophisticated interface are required to perform functions that could be integrated into a single unit. Both large and small customers are likely to prefer lower cost, functionally integrated equipment rather than more expensive equipment that conforms to the ITU-T reference model.

## 7.5 DEPLOYMENT OF ATM ACCESS NETWORKS

ATM is likely to be used first by business customers who already have high-speed leased circuits or ISDN services, because these customers are familiar with sophisticated digital services and are more likely to be prepared to pay the high initial charges. Although residential services such as VoD can also be supported by an ATM fabric in the access network, the use of ATM here is likely to be delayed because the revenues generated are unlikely to support the investment required until the demand increases and the equipment costs decrease.

Business customers are likely to require symmetric, bidirectional ATM transport, because they will often need to communicate with other's businesses on a peer-to-peer basis. In some cases, the communications will not be symmetric, since the business customers may need to access databases. Business customers who provide database or video services may also have asymmetrical transport requirements, but with the major direction of transport being outgoing rather than incoming. Residential customers are also likely to have asymmetrical transport requirements due to the demand for entertainment services.

## 7.6 SUMMARY

ATM provides a common fabric for telecommunications networks that is largely independent of the physical layer that supports it.

Within an access network, ATM can be used both over copper pairs and over optical fiber. The use of ATM for radio access networks is less appropriate, since the limited bandwidth of radio systems restricts their use for broadband. Point-to-point bidirectional radio links are better suited for ATM, because they can support higher bandwidth applications, since their radio spectrum can be reused in different locations. If ATM is used as the transport layer for a VoD service over copper pairs, then migration of the service onto optical fiber is simplified, since the higher architectural layers do not need to be changed.

ATM cells provide a natural basis for the multiplexing of traffic from the remote ends of an ATM PON. It is also natural to add an additional label to the normal 53-byte ATM cells to differentiate between traffic and control cells, and to help with point-to-multipoint multiplexing, although an additional label is not strictly necessary. Ranging and leveling are necessary if TDMA is used on the upstream transmission to compensate for the different transmission delays and attenuations of the different remote ends. There are advantages to using a hybrid ranging scheme with pseudorandom low-resolution ranging and a time domain window for high-resolution fine-tuning, since this can be faster and the time domain window can be short. Continuous transmission in the downstream direction simplifies the synchronization of the remote ends.

Capacity allocation in the downstream direction is easy, since it can be achieved simply by the appropriate addressing of ATM cells. In the upstream direction the problem is more difficult, since capacity has to be allocated to the remote ends dynamically according to the demand, because some time slots in the upstream direction might be used by any of the remote ends if TDMA transmission is used in the upstream direction.

It is not possible to support the signaling for existing narrowband services using broadband Q.2931 signaling for the same reasons that protocols in addition to narrowband Q.931 were needed to support PSTN and basic rate ISDN on V5 interfaces. There are also problems with the delay introduced to fill ATM cells for narrowband voice traffic, and it may be easier for voice traffic to be split off into a traditional narrowband part of the telecommunications network and not handled by the ATM network.

There are problems due to the imbalance between the large number of addresses allocated to user ports and the small amount of additional address space provided at an ATM network-network interface. This imbalance makes it difficult to avoid VP switching in the access network and VC switching may also be needed. Reservations have been expressed about the ITU-T reference model for ATM accesses, and customers may prefer to buy lower cost equipment rather than conform to the ITU-T model.

ATM is likely to be used first for business customers, who are more likely to require the same bandwidth in both directions than residential customers, who may require more downstream bandwidth than upstream bandwidth. Business customers may even need more upstream bandwidth than downstream bandwidth, particularly if they act as service providers.

## Selected Bibliography

Angelopoulos, J. D., I. S. Venieris, and G. I. Stassinopoulos, "A TDMA Based Access Control Scheme for APONs," *J. Lightwave Technology*, Vol. 11, No. 5-6, May–June 1993, pp. 1095–1103.

Chen, T. M., and. S. S. Liu, "Management and Control Functions in ATM Switching Systems," *IEEE Network*, Vol. 8, No. 4, July–August 1994, pp. 27–40.

de Pryker, M., *Asynchronous Transfer Mode Solution for Broadband ISDN*, New York: Ellis Horwood, 1993.

Farkouh, S. C., "Managing ATM-Based Broadband Networks," *IEEE Communications Magazine*, Vol. 31, No. 5, May 1993, pp. 82–86.

Fischer, W., E. Wallmeier, T. Worster, S. P. Davis, and A. Hayter, "Data Communications Using ATM: Architectures, Protocols, and Resource Management," *IEEE Communications Magazine*, Vol. 32, No. 8, August 1994, pp. 24–33.

McDysan, D. E., and D. L. Spohn, *ATM Theory and Application*, New York: McGraw-Hill, 1995.

Tusoboi, T., Y. Maeda, K. Hayashi, and K. Kikuchi, "Deployment of ATM Subscriber Line Systems," *IEEE J. Selected Areas in Communications*, Vol. 10, No. 9, December 1992, pp. 1448–1458.

Verbiest, W., G. Van der Plas, and D. J. G. Mestdagh, "FITL and B-ISDN: A Marriage With a Future," *IEEE Communications Magazine*, Vol. 31, No. 6, June 1993, pp. 60–66.

Walters, S. M., D. S. Burpee, and G. H. Dobrowski, "Evolution of Fiber Access Systems to ATM Broadband Networking," *Proc. IEEE*, Vol. 81, No. 11, November 1993, pp. 1588–1593.

# 8 Services Supported by Access Networks

"So desperately hungry for material that they're scraping the top of the barrell."

—Gore Vidal

## 8.1 INTRODUCTION

The emergence of new telecommunications services may be similar to the emergence of applications software after the personal computer became established. In both cases a basic platform is required and the precise applications are not easy to predict in advance. However, the development of telecommunications services has lagged behind the development of personal computer applications software, and advances such as fax and e-mail sometimes seem to have occurred despite the activities of the international telecommunications fraternity.

The culprit responsible for the delay in the development of advanced telecommunications services may be the fallacy that these require centralized control by the telecommunications operators. Many of the services that were originally intended to be provided by telecommunications operators through new digital exchanges are now provided by the telephone handsets themselves. Even a feature such as call forwarding, which requires interaction with the exchange, can be replaced by an appropriate message left on an answering machine, since this fulfills the same functional requirement and is simpler to implement.

The promise of centrally controlled telecommunications services appears to have been as fruitful as the promise of centrally controlled communist economies. The growth of new services has been most vigorous at the periphery of the telecommunications network, especially where this has not been subject to bureaucratic restraint. This growth has involved using modems to overcome the limitations of conventional analog telephony rather than

ISDN, which has received much more attention but which has at times deserved its description as "Innovation Subscribers Don't Need" and as "It Still Does Nothing."

Despite the poor start, several aspects of future telecommunications services are now becoming clear. The most popular residential service appears to be for video entertainment, with interactive program guides and on-demand selection from extensive libraries. There also seems to be a demand for home-shopping and financial services. The most socially significant aspects may be telecommuting and telepresence, which allow people to work from home, and the Internet, which allows people access to information and services anywhere on the globe. Simple global communications may be better served by e-mail and fax than by voice or video telephony, since the windows for real-time global communication is limited by differences in time zones. E-mail in particular is simple and fast and avoids the frustrations of playing telephone tag when the called party is not available. Telecommuting and telepresence in themselves will not be considered further, however, because they can be considered as ways of using other services and not services in their own right.

It has become clear that directionally symmetrical bandwidth, which is typical of conventional telecommunications, is not appropriate for many of the new services. The downstream data rate for access to films, databases, libraries, and games vastly exceeds the upstream rate, since the upstream transmission is mainly used to control the incoming information. Many of these new telecommunications applications will be critically dependent on connection charges. For example, it is likely that there will be a significant demand for multiplayer video games, since games can be more interesting when played against real opponents and in real time, but the attraction of multiplayer games will be decreased if the cost of the communications is high.

The availability of appropriate communications software is increasingly a critical factor in the development of application services, even for the existing analog network with modem communications links. With the present communications technology, suitable software would allow lectures to be multicast on the Internet. The development of software for sophisticated browsers and hypertext links has revolutionized the use of the Internet, and the subsequent increase in traffic is making further evolution of the Internet necessary.

The similarity drawn between application software for personal computers and application services for telecommunications networks may be more than just an analogy. The two may be one and the same, since the application services may be the capabilities that are supported by the application software running on users' computers at the periphery of the telecommunications network. Furthermore, the application software itself may be developed and distributed using telecommunications networks working in cooperation with computer networks.

## 8.2 SIMPLE VIDEO SERVICES

There appears to be a growing dissatisfaction with the services provided by CATV. The popularity of video rentals indicates both that customers are interested in a wider range of services and in a greater control over their scheduling than is supported by CATV. Future video services are unlikely to be based on conventional terrestrial broadcasting, since the radio spectrum available is limited and better use can be made of this scarce resource. The future of video seems to be with satellite systems or with access networks that have a fixed infrastructure, rather than conventional broadcasting or conventional noninteractive cable distribution.

### 8.2.1 Video-on-Demand Service

VoD is, in effect, an electronic video rental service, but with the advantages of on-screen ordering, immediate availability, familiar VCR-like controls, and no difficulties due to popular titles being out on loan. The use of the MPEG compression standards and ADSL transmission allows VCR-quality video to be provided on most existing copper pairs. Higher quality service can be supported on shorter lines or if the transmission link is supplemented by active nodes. The use of copper pairs for VoD may be an effective introductory strategy, since it makes use of the existing infrastructure. It is unlikely to be an appropriate long-term solution because it does not support a good quality of transmission on all lines without remote active nodes, which are expensive to maintain.

The nature of the video transmission for VoD need not be limited to library material. The video image could be created by the users themselves in various ways. For example, the displays of a multiplayer video game could be transmitted to the players, who then use their low-rate upstream control channels to influence the development of the game. The cost of playing such a game should be comparable to the cost of watching a video from a library, since the bandwidth requirements are similar and the cost of computer processing power is constantly decreasing. However, it should be noted that there are much tighter restrictions on response times for these video games than for true VoD if the feel of existing video games is to be preserved. The technology can also be used for the staggered broadcasting of conventional TV programs so that viewers can watch their favorite shows when it is convenient rather than when they are broadcast and without the trouble of arranging to record them.

### 8.2.2 Video-on-Demand Architectures

Even if the transmission technique does not limit the bandwidth of a service, the physical medium does. This may be more of a problem for the storage medium in the video server than in the access network, since the access network needs to be designed so that users do not interfere with each other, whereas several users may attempt to access the same data in the video server at only slightly different times. If users have truly independent access to the video servers, then potentially large amounts of additional data storage will be required. Less storage and lower bandwidth is needed if users access the same data simultaneously than if their accesses are truly independent, since data can then be multicast to several users.

Bandwidth requirements can be further reduced if a single common VoD server is supplemented by a number of local video buffers, since this reduces the number of independent links to the VoD server. The use of supplementary video buffers may be essential, since otherwise gigabits per second of transmission would be required at the server for it to support about a thousand users. Several users can access the same video transmission in the local video buffer by tapping the buffer at different points. Popular videos are likely to be simultaneously accessed by a number of users, since the number of currently popular videos is likely to be much less than the number of users. The larger the local buffer, the greater the probability that a later user can tap a delayed version of the transmission to the initial user without the need for the establishment of a second link to the VoD server. It would also be possible to reduce the bandwidth to the VoD server by keeping copies of popular videos permanently at the local video buffers. Such a local video buffer would essentially be a small VoD server, which allowed the backbone traffic to be decreased at the cost of additional capital investment in VoD equipment.

In the traditional view of services, the common VoD server and its satellite buffers would be owned by the national service operator. There is no reason why third parties, independent of the telecommunications operator and the customer, could not be involved here. This diversity could lead to greater variety and lower costs to the customer. In this case, the local buffers could provide locally popular videos directly and also act as brokers to provide less popular videos from the remote sources. These sources could either be a large "warehouse" VoD server or other local buffers specializing in particular stocks. Brokers could also provide specialized services. For example, a broker in a particular country could specialize, perhaps, in adding automatic subtitles to imported films in a foreign language.

### 8.2.3 Video Over ATM

ATM has been proposed as the underlying infrastructure for VoD by some proponents because of its flexibility. This can be described as making the service future-proof, since ATM does not restrict the bandwidth. On the other hand, it can be considered as an admission of uncertainty about the nature of a VoD service, since the cost of using ATM could be eliminated if the nature of the service were clear. An optimist might view the use of ATM as inspired. A pessimist might compare it to a drowning man clutching at a straw.

One of the advantages of ATM is that it provides a common bearer that is independent of the physical medium used in the access network. This allows the services to decouple from the physical medium, permitting the physical medium to be changed without needing to modify the service. It is not clear that this is really a justification for ATM, since the basic PSTN service can already be carried on a number of different media without this fuss. Where ATM may be able to give added value is in providing a common infrastructure for several different services and in simplifying the evolution to higher rate broadband services.

A further advantage of ATM is that it allows the bandwidth overhead required for signaling to be minimized, since it is not necessary to dedicate a fixed-capacity channel for signaling that is normally inactive. This becomes more important if a large number of low-bit-rate services need to be simultaneously supported. A more conventional approach of combining the signaling channels might introduce a higher level of interaction between services if the signaling requirements of one of the services suddenly increases. However, ATM can also suffer from interaction problems, since it makes use of the statistical multiplexing of traffic.

### 8.2.4 Satellites vs. Access Networks

Access networks appear to have an advantage over satellite video broadcasting because of the high-bandwidth capability of optical fiber and because access networks provide a separate path to each customer. This advantage is not as great as it may initially seem because of the large possible number of geostationary satellites, giving satellites the potential to support a wide choice. Satellite broadcasting also avoids the need for an expensive fiber infrastructure. The infrastructure costs are less if an existing copper infrastructure is used for the access network, but the equipment costs would also have to be comparable to a satellite receiver.

Access networks can serve a different market segment than satellite broadcasting because they can support interactive services tailored to the individual customer, rather than actual broadcasting. Satellites are not well suited to tailored services, since satellites are more appropriate for multicast or broadcast operation, but this covers a large segment of the entertainment market.

Satellites can also support a form of VoD through staggered broadcasting of the same video on a number of different channels. This is a lower quality of service than a true on-demand service, which can be offered on an access network, but may be adequate for many customers and may be sufficient for the distribution of current "top 10" video choices. It is also not clear that the revenues that can be generated by a true VoD service are sufficient to support the installation of a widespread access network infrastructure, since people are unlikely to watch many videos in a week and the cost of the service cannot be significantly higher than the cost of physical rental. Concerns about the cost sensitivity of the demand for VoD has led to the coining of the term "video-no-demand."

Access networks may have an advantage over satellites for high-definition television (HDTV) and future digital video technology, since it is easier to upgrade terrestrial equipment than orbiting satellites or to launch new ones. In the long term, the emphasis within access networks is likely to be on fiber or fiber/coax systems, since the bandwidth available on twisted pair or radio systems is limited. In the shorter term, there may be widespread use of ADSL technology, since it can support video of a moderately good quality.

Satellites are quite unsuitable for video telephony, since it requires a high-capacity bidirectional link. It is possible to support video telephony over a narrowband-ISDN link, but a higher bandwidth seems necessary for a good-quality service. However, it seems that the attraction of video telephony has been overrated in comparison to other video services, since customers show a reluctance towards it for apparently psychological reasons.

## 8.3 MORE SOPHISTICATED SERVICES

Although simple video services, such as video telephony and VoD, are relatively easy to imagine, in the longer term more sophisticated interactive services are likely to emerge. It is less easy to predict the nature of these more advanced services, since the demands they will meet will also depend on the development of technologies outside of telecommunications.

### 8.3.1 Tailored Advertising and Telepurchasing

The range of items that may be remotely purchased is not simply limited to those normally labeled as "home shopping." The early forms of home shopping simply consist of broadcast advertisements combined with ordering by telephone and credit card and mail delivery, with the communications typically over a cable TV network.

A more effective approach is to tailor the advertisements to particular profiles of customers. This requires a means of narrowcasting advertisements to the appropriate groups. Even without telephone ordering and mail delivery, advertising is more effective if it is directed to particular audience, since there will be a higher level of interest. This is already done in a limited way by scheduling advertising to match the program content, but a separate channel to each customer allows this to be made more personal. In more sophisticated home shopping, the crude mass broadcasting or narrowcasting of advertisements is replaced by interactive access, either through the selection of broadcast material or through individual video browsing of virtual shops. In the initial stage, an electronic notice board could be used, possibly organized into categories, to provide an electronic classified advertising service. Typically, this could also eliminate the need for a separate telephone call to order the items, since the order or reply could be generated immediately.

This form of home shopping may well be overrated. It may be more effective for existing mail-order companies to send out their magazines in the form of CDs, since many homes already possess a computer with CD capability. This would reduce the publication costs of the mail-order companies and avoid increasing the communication charges experienced by their customers.

The next level of sophistication, and perhaps the first viable form of telepurchasing, involves the elimination of mail delivery. This will still be required for physical goods, since technology capable of teleporting these goods to their destination is not yet feasible. The elimination of mail delivery becomes feasible in certain cases in which the goods to be delivered consist of information. A VoD service is effectively a form of purchased information, although it is not normally considered as such because the information will not normally be stored. Music could also be delivered in this way, and it may be more appropriate for music to be stored and replayed later, since music is typically replayed more often than videos. In both cases, however, it is easier to transmit data for storage and later replay than to transmit it for real-time display, since the constraints on delay and jitter are less.

Software is "replayed" as much as or even more than music, and downloadable shareware is already available on demand via the Internet. This raises the issue of software piracy, and more generally that of data piracy, since the ease of copying and transmitting data undermines copyright and ownership legislation. It is not clear how this issue can be resolved, since the information must be converted into a usable form at the final destination regardless of the type of protection used.

Data communications both aggravates the problem of data piracy and may provide a potential solution. It aggravates the problem by facilitating the distribution of pirated data. The solution to this may be ironic in that the source of the problem could be the source of the solution. It may be better to supply information readily and at low cost through telepurchasing rather than to attempt to protect it. This may be successful because high prices and protection mechanisms serve to create the environment in which a pirate can operate. Prohibition laws for information will probably be as successful as prohibition laws for alcohol, and in the long term may go the same way. It is probably more sensible to rely on natural honesty, since people are content to act within the law unless the return is high and the chance of getting caught is small.

### 8.3.2 Internet Services

The basic services supported by Internet are e-mail between users, remote login to other computers, and the transferring of files to and from servers. Typically, the Internet allows access to global data communications for the cost of a local telephone call to a server, and the cost is even less if the user is directly connected to the server. E-mail is also available to a large number of people who do not have full Internet access and is widely used because it is fast and allows written text or suitably encoded documents to be transferred between different computers. Users who have this form of limited access are sometimes described as being on an *outernet*, and the combination of Internet and the outernets is sometimes referred to as the *Net*. More recently, the use of the Internet has entered a new phase with the development of hypertext links and the creation of the World-Wide Web (WWW) and sophisticated browsers, which make the Internet more accessible to nonexperts.

The growth of Internet traffic has been exponential. In fact, this growth has exceeded the rate at which new resources have been added. The outstripping of resources indicates that the exponential growth of traffic must begin to slow down. There are signs, however, that this might entail a change in the rate of exponential growth rather than a change in the nature of the growth. This is confirmed by measurements that indicate that different services have different exponential growth rates. However, the growth of the browsing traffic on the WWW has been so great that it has been called the *killer Internet application*.

The growth of Internet traffic has been partially due to new users discovering the Internet. The growth has also been driven by existing users making greater use of the Internet as their work patterns change. These factors are amplified by the increasing ease of use made possible by new user-friendly tools. It is likely that the nature of the Internet will change as a result of its increasing popularity, since the cost of its operation is often not reflected in the charges for its use.

It should be recognized that for many people, the new telecommunications services have already arrived on the Internet. Traditional communications operators who wish to remain in business may need to wake up to this.

### 8.3.3 Distance Learning

Distance learning is a general term to cover the educational use of telecommuting. It typically has an asymmetrical transmission requirement similar to that of VoD, since more information needs to be transmitted to students than is received from them. In its crudest form, distance learning need not be more than VoD with an educational content. A more sophisticated form would allow the real-time multicasting of lectures and give students the ability to ask questions through voice conferencing as if they were physically present in the lecture hall.

Depending on the implementation, it may be better if questions are typed rather than spoken, since audio conferencing presents difficulties for computer networks. Written interactive communication has the additional advantages of being easier to control and producing a record for later review.

The video bandwidth required for a lecture or presentation is less than that required for a VoD service, since the information presented does not change rapidly. A rule of thumb for presenters is to allow two minutes per slide for talking through and indicating the important aspects with a pointer. The bandwidth for this is at least two orders of magnitude less than that for normal video. These data rates can be easily handled by a narrowband-ISDN link. The underlying technology for this narrowcasting of lectures has already been tested using the Internet, and the more sophisticated types of modems would even allow distance learning to be supported through Internet access over existing analog lines.

A critical factor here, as for many applications, is likely to be the connection charge. The multicasting of lectures could rapidly flourish if the cost is equivalent to a local PSTN call. In the longer term, it is not clear that colleges and universities need to continue in their present form, although psychological factors, such as snobbery, may play a significant role.

### 8.3.4 Video Telephony and Multimedia Services

The nature of personal computer applications is changing due to the advent of low-cost compression and expansion chips for audio and video data and the increasing capacity of hard disk memories. The use of compression techniques is widespread in software distribution, where application software is commonly compressed for distribution and then expanded to executable code when installed; even data files, such as documents, are often compressed before they are electronically distributed. However, the compression of video and audio files is different because it adds both sound and a dynamic dimension to traditional text or diagrams.

The presence of audio and video on personal computers raises the question of what their relationship will be to more traditional communications equipment such as televisions and telephone sets. It is unlikely that personal computers will supersede conventional telephone sets because in many cases the cost of the personal computer cannot be justified. The situation for video telephony is less clear, particularly in an office environment where computers are more common. It is likely that using the audio and video capabilities of personal computers to support video telephony will be more attractive than using a separate expensive and nonintegrated video-telephone set. The advantages of this approach are increased due to the problems of communicating across different time zones and when the called party is unavailable, because this approach is also suited to the transmission of a compressed recorded video message.

The residential situation is more complex because of the requirement for entertainment services that may require better downstream quality that that necessary for purely office communications. Video telephony could make use of televisions for display and use camcorder cameras for transmission. Although this appears feasible, the personal computer route may still be more attractive. Personal computers are increasingly common within homes, and their use for video telephony in business could lead to low-cost consumer versions. Perhaps more significantly, a video telephone should be able to operate independently of the residential entertainment system, because other members of the family will not wish their entertainment to be disrupted and because privacy is often required. The difference between the office and the residential requirements should also be further reduced if telecommuting becomes more commonplace.

### 8.3.5 LAN Interconnection and Emulation

Computer networks often consist of LANs, which are interconnected by gateways and routers to form *wide-area networks* (WAN). To be successful in this area of communications, access networks must be capable of providing gateway and routing services and of emulating LANs so that computer equipment can be transparently linked to them without the conventional telecommunications being visible. LAN emulation is achieved by creating a virtual shared medium that can be used as if it were a shared physical medium such as an Ethernet cable. To support the increasing demand for Internet communications on LANs, a connectionless datagram service, which is independent of the physical and logical paths between its end points, must also be supported.

A characteristic attribute of LANs is that they use broadcast communications. Each station on a LAN typically receives all of the packets transmitted by the other stations and ignores those packets that are not addressed to it. This is a natural form of operation for both optical and wireless access networks, since broadcast transmission is used for both. It also fits well with ADSL operation, since LAN stations typically transmit less than they receive. The PON architecture of access networks make them particularly well suited to the implementation of LAN bridges. In the simplest form, these bridges operate by listening on one LAN and transmitting on all other related LANs. PONs are well suited to this because of their point-to-multipoint structure.

## 8.4 TELEACTION SERVICES

In contrast to the more glamorous high-bit-rate services, such as VoD and LAN emulation, the more immediate new telecommunication services may make use of low bit rates to carry out some simple action such as data collection. Services supporting simple remote transactions and control are referred to collectively as *teleaction services*. Interest in teleaction services is consistently demonstrated by potential users, unlike the situation for many proposed services, and the challenge here is to provide them cost-effectively. For credit card validation and automatic teller machines, this challenge has already been met.

### 8.4.1 Alarm Services

Alarm services can be characterized by analogy with the emergency services for police, fire, and ambulance. The alarm service for burglary detection differs from those for fire and medical emergencies, since burglars may disable the alarm. Fire and medical emergencies do not attempt to disable the service supporting the alarm. An alarm service must raise an alarm if communication is lost or interfered with, since it should detect tampering by burglars. This is also useful for fire and medical emergencies, since fire could disable communications and a loss of communications could jeopardize life in a medical emergency.

One of the greatest problems with fire and medical emergencies is the occurrence of false alarms. In particular, some of the users of medical alarms are tempted to test them without proper appreciation of the consequences. This can be countered by a confirmation sequence, for example, calling the user to confirm that the alarm is genuine. Unfortunately, this introduces a delay that could be fatal for fire and medical emergencies and may be counterproductive if the time delay allows an intruder to take the user captive.

Burglars may also be tempted to create a pattern of false alarms to buy extra time when the genuine alarm occurs. In extreme cases they may even attack the communications network itself so that their particular target is not identifiable. Alarm communications can be made more immune to tampering if radio communications with interference protection is used, and radio systems are more appropriate for mobile medical alarms and ease of installation.

### 8.4.2 Yellow Pages Services

Yellow pages services are likely to have the highest bandwidth requirements of all teleaction services, since they involve the presentation of information to customers. They differ from the conventional approach to home shopping, which has its roots in broadcast video. Electronic yellow pages services are likely to be more detailed than the conventional paper-based approach, since they can provide more information about specific products. This type of communication also enables the introduction of inventoryless retailing, where the retailer can process the order and then place a batch of orders with the wholesaler's warehouse, eliminating the need for the retailer to keep a physical inventory.

### 8.4.3 Services for Utilities

The teleaction services required for utilities (gas, water, electricity, and telecommunications) need to be considered differently for other products that are sold to customers, since they require lower bandwidths and support different functions. The most obvious function is remote meter reading, since this avoids the expense of sending a human being to do the task. It also has the advantage of supporting direct billing. This service encourages the prompt paying of bills and the interest so saved is a significant incentive to the utility companies. Direct electronic billing can also be cheaper than conventional billing, since there is no need to send bills by mail. There are further simplifications possible if payment is also electronic. It is ironic, and very human, that telecommunications operators have not taken the initiative in this area to reduce their own costs in billing their customers.

There are also advantages to demand management for utilities, especially for electricity, since electricity is difficult to store. Demand management can be achieved either by direct control of consumption by teleaction or by indirect control through the notification of tariff changes. The latter approach could be integrated with an intelligent home controller that could shop around for the best deal at any time.

### 8.4.4 Intelligent Remote Equipment

Direct demand management and home controllers both rely on the existence of appliances that can be remotely controlled. Once these exist, they enable the creation of further home control services that can be activated remotely. These can range from turning up the heating before the owners return home to defrosting and microwaving food for their dinner.

Food vending machines can also benefit from remote telemetry. It may not be possible to eliminate the vending of hot, sweet, milky water instead of coffee, but its occurrence may be reduced if the inventory is monitored remotely. When problems do occur, it may also be possible to use teleaction to register a complaint. Faults could also be communicated to a remote maintenance facility.

### 8.4.5 Implementation Issues

Many of the teleaction services identified require a small number of bytes to be transferred rapidly between terminals. The PSTN network is capable of performing this task, but the PSTN network is better suited to long-duration analog calls. The packet and frame-relay capabilities of narrowband-ISDN are more suitable, but an ISDN terminal can cost more than using a modem on PSTN. What is required to meet the proven demand for teleaction services is a simple datagram capability with a low implementation cost. Unfortunately, the great minds of the international telecommunications fraternity appear to be too finely honed to be able to handle such a crude request.

The rapidity of the deployment of teleaction services depends largely on the availability of low-cost communications capabilities that can be readily combined with other equipment that provides the primary function. Potential users are likely to be concerned about losing control of their equipment and appliances, about the possible invasion of their privacy, and about having intelligent equipment that can generate large phone bills. If these problems can be overcome, then the services are likely to grow rapidly despite any shortcomings of the telecommunications infrastructure.

## 8.5 SOCIAL AND PSYCHOLOGICAL FACTORS

In many cases there is no technological or economic barrier to the introduction of new telecommunications services, but the service is not successful for social or psychological reasons. Often this is not understood by those who have helped to develop the service and who often have a background in hard science. Unfortunately, the social and psychological factors that hinder this understanding also lead to repeated forms of displacement activity, where the technical issues are tackled and retackled instead of the required nontechnical solution being sought.

### 8.5.1 Video Services

The significance of psychological factors in restricting the emergence of new video services has long been overlooked. In theory, video-conferencing should have been extremely popular as an alternative to long-distance travel, since it can cost less. In practice, video-conferencing has been more effective if the distances are short. In part this is due to the benefits of traveling long distances on business and the opportunities for hospitality and socialization that travel permits. If the distances are short, then the psychological disincentives are eliminated and there is less reluctance to try out the new technology, because it is easier to make alternative arrangements if it fails.

Likewise there have been repeated trials of video telephony that have not succeeded in establishing this service. Part of the reason for this is the preference for purely vocal communication in many cases. The visual component in communications can be felt to be a distraction, and many people feel they communicate better when they can concentrate on the vocal aspects without worrying about their visual impact. The logical development of this is the use of e-mail that also removes the nonverbal vocal component, such as accent or tone of voice, and allows pure communication of ideas untarnished by nonverbal distractions. Video telephony is also perceived as more invasive of privacy than voice communication. Video telephony is much more likely to become established if it becomes a social stigma to only have a voice telephone.

Psychological factors have also been ignored in the initial analysis of the required quality for distributed video services, which has suggested that this need only be equivalent to that produced by customers' VCRs. This suggestion may be somewhat naive, since this could create a poor market image of the new services, since their quality would be lower than that of existing broadcast services. It has also been suggested that customers would be content to use DTMF (dual tone multifrequency) signaling or telephony-based signaling, such as voice recognition, to control video services. There appears to be an element of wishful thinking in this, since customers would probably really prefer to use remote control through infrared wands with which they are familiar, rather than a more cumbersome approach that fits better with traditional telecommunications.

### 8.5.2 Telemedicine

There has also been a naiveté among certain of the advocates of new services that borders on certifiable insanity. Advocates of telemedicine are perhaps the most infected. Medical doctors have been offended by suggestions that it is sufficient for them to see a video of a patient in order to diagnose the problem. It is unlikely that doctors have been reassured by the often related suggestions that the doctors themselves could be replaced by expert systems. In fact, the more evidence that is accumulated about how this could be done, the greater the objections by doctors are likely to be.

The greatest failing of advocates of telemedicine is probably the inability to learn from existing applications. The Australian "flying doctor" service is an excellent example of telemedicine. This service would not necessarily be significantly improved by video transmission, because a detailed diagnosis is generally not what is needed. What is required is the ability to make an accurate decision about whether an ambulance plane should be sent to the patient. Likewise, in other situations telecommunications can make a greater contribution by enabling a fast and appropriate response to emergencies, rather than by reducing the human element in medical care.

Perhaps the greatest case for telemedicine may be that the remote analysis of the publications of some of its more extreme advocates permits a diagnosis to be made of their mental state and the social and psychological factors that have influenced them, possibly, since childhood.

### 8.5.3 Electronic Communities

One of the greatest incentives to use new services is also social or psychological. It is the desire to belong to a community that shares similar cultural values.

Bulletin boards and e-mail groups enable the creation of a number of virtual Bloomsbury sets to talk and exchange ideas about common areas of interest throughout the world. This incentive is especially strong among professional groups with esoteric specializations.

A more active form of communication may act to counter the "couch potato" response to broadcast communications. This may be overly optimistic, since, before the development of game shows, television was once considered to be the medium by which the mass of humanity would gain access to culture and education. The incentive to form groups with similar cultural values is also shared by cabals, cliques, and conspiratorial cells, and it may be worth noting that electronic bulletin boards have also been used for the coordination of international activities by extreme neo-Fascist groups. The freedom of expression that already exists in common interest groups using electronic communications can be described, according to one's political outlook, as either dangerously subversive or refreshingly enlightened. It may be both.

## 8.6 SUMMARY

There is a similarity between the development of telecommunications services and software applications for computers, but the development of telecommunications services seems to have been delayed by the centralized planning of the networks, and the directionally symmetric bandwidths of conventional telecommunications are not appropriate for many new services.

The future of simple video services seems to lie with access networks and satellite broadcasting. Satellite broadcasting is better suited to mass communications without interactive control, but the possibility of large numbers of satellites and closely staggered transmission challenges the use of access networks for more specialized video services. VoD services may be best initially delivered on existing copper pairs, but in the longer term a fiber infrastructure will be needed to provide the bandwidth and individual connections. These individual connections may need to be buffered, possibly in the access network, since otherwise VoD servers may need unacceptably high bandwidths because of the large amount of separate accesses to a small number of popular videos. The use of ATM in the access network may be better justified in terms of providing a common infrastructure for different services and simplifying upgrades rather than in terms of media independence, which has already been achieved for telecommunications services.

More sophisticated services are likely to include teleshopping, especially for downloadable data, and tailored advertising. There will also be a significant use of the Internet, although the nature of the Internet is likely to change due to the escalating demand. Conventional universities may be superseded by telecommuting to distance learning centers, but this is highly dependent on the connection tariffs. The increasing use of personal computers is likely to lead to their adaptation to support multimedia services and video telephony, initially in the office environment but afterwards in the home. LAN interconnection services and LAN emulation will be needed to link computer networks, and PON and ADSL technologies have natural advantages for LAN broadcasting and bridging.

There is considerable demand at the other end of the bandwidth spectrum for low-data-rate control and transactions, known as teleaction services. These include alarm services, yellow pages services, and utility services, including control, meter reading, and billing. The development of low-cost intelligent equipment in the customers' premises will spur the development of teleaction services, so long as the charges for low-bit-rate communications are sufficiently low and unauthorized access is difficult.

The greatest barrier to the development of access networks to support these new services is likely to be social and psychological rather than technical. Video-conferencing lacks the social advantages often associated with business travel, video telephony is more psychologically intrusive and distracting than voice telephony or e-mail, and proposals for telemedicine have alienated doctors and failed to learn from the lessons of the successful "flying doctor" service. Fortunately, the development of electronic communities and peer groups, which has already started through the Internet, may provide a positive social influence encouraging the development of new services and the access infrastructure needed to support them.

## Selected Bibliography


Brunet, C. J., "Hybridizing the Local Loop," *IEEE Spectrum*, Vol. 31, No. 6, June 1994, pp. 28–32.

Burpee, D. S., and P. W. Shumate, Jr., "Emerging Residential Broadband Telecommunications" *Proc. IEEE*, Vol. 82, No. 4, April 1994, pp. 604–614.

Chao, H. J., D. Ghosal, D. Saha, and S. K. Tripathi, "IP on ATM Local Area Networks," *IEEE Communications Magazine*, Vol. 32, No. 8, August 1994, pp. 52–59.

Crutcher, L., and J. Grinham, "The Networked Video Jukebox," *IEEE Trans. Circuits and Systems for Video Technology*, Vol. 4, No. 2, April 1994, pp. 105–120.

Deloddere, D., W. Verbiest, and H. Verhille, "Interactive Video on Demand," *IEEE Communications Magazine*, Vol. 32, No. 5, May 1994, pp. 82–88.

Dyke, P. J., and D. B. Waters, "A Review of the Technical Options for Evolving FITL to Support Small Business and Residential Services," *J. Lightwave Technology*, Vol. 12, No. 2, February 1994, pp. 376–381.

Macedonia, M. R., and D. P. Brutzman, "MBone Provides Audio and Video Across the Internet," *Computer*, Vol. 27, No. 4, April 1994, pp. 30–36.

Paxson, V., "Growth Trends in Wide-Area TCP Connections," *IEEE Network*, Vol. 8, No. 4, July–August 1994, pp. 8–17.

Scott, H. A., "Teleaction Services: An Overview," *IEEE Communications Magazine*, Vol. 32, No. 6, June 1994, pp. 50–53.

Sutherland, J., and L. Litteral, "Residential Video Services," *IEEE Communications Magazine*, Vol. 30, No. 7, July 1992, pp. 36–41.

# 9 V5 Interfaces and Architecture

"There is no need for housekeeping at all, since after four years the mess doesn't get any worse."

—adapted from Quentin Crisp

The V5 interface is the interface between an access network and a host exchange for the support of narrowband telecommunications services. It has two forms, V5.1 and V5.2, and the more sophisticated V5.2 form supports the concentration of traffic. The apparent complexity of the V5 interface is a result of meeting a number of simple basic requirements, and the avoidance of unnecessary functionality has been a driving consideration. One consequence of this is that in-band tones are passed transparently across the V5 interface and the responsibility for their generation and detection remains with the host exchange, not the access network.

In this chapter the architectural model associated with the V5 interface is described, together with the physical structure of the interfaces and the services they support. Subsequent chapters describe the general multiplexing and format of V5 messages, the multiplexing and handling of ISDN signaling, and the various V5-specific protocols. The chapters on the V5 control protocol and the protocol for PSTN signaling are relevant to both V5.1 and V5.2 interfaces. The later chapters on the BCC protocol, the link control protocol, and the protection protocol are only relevant to V5.2 interfaces. Broadband services are supported by VB5 interfaces, which are introduced more fully towards the end of the book.

The V5 interface is not limited to any specific access technology or medium, although much of the motivation for its development was the anticipated deployment of optical access networks. There has been considerable interest in its use for radio access networks, which, with hindsight, should have been obvious because of the technical and commercial developments in mobile radio access. Also with hindsight, the potential use of the V5 interface for interconnection between telecommunications networks belonging to different operators should have been more obvious.

Because it ignores the details of access technology, the model of an access network appropriate for a V5 interface can appear out of proportion to the access network as a whole. From the point of view of a V5 interface, an access network is a black box with borders but no internal structure. Features that are optional and viewed as just outside the borders are more important than those that are essential but not relevant to the V5 interface. For example, the optional transmission systems at the borders are of more concern than the internal transmission system of the access network.

To understand the V5 interface, it is necessary to first understand the nature of the boundaries of the access network. These include the physical links to the exchange, the physical links to the remotely located user ports, and the services supported at the user ports. Once the nature of the boundaries are clear, the details of the V5 protocols and the ways they are multiplexed together can be examined in context.

## 9.1 THE V5 ACCESS MODEL

The architectural model for the V5 specification differs from other models sometimes used for access networks because it is focused on the aspects relevant to the V5 interfaces and ignores those details that are only relevant to the specific access technologies. The basic model is shown in Figure 9.1.

It is necessary to distinguish between the *local line distribution network* (LLDN), which stretches from the host exchange in the core network to the *customer premises equipment* (CPE), and the access network as it is defined for the V5 interface. The difference between the two is that the LLDN also includes any *feeder transmission systems* (FTS) and *remote digital sections* (rDS) if they are present.

An FTS allows the headend of an access network to be located remotely from its host exchange. The FTS could take the form of an SDH ring with

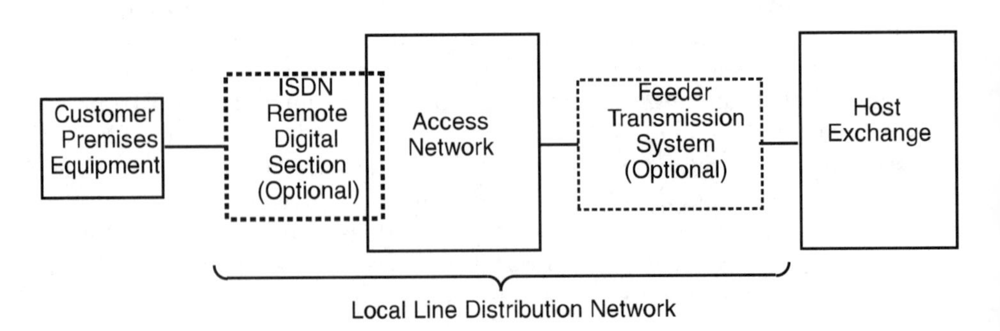


**Figure 9.1** The V5 access model.

add-drop multiplexers at the host exchange (or host exchanges) and also at the locations of the headend of the access network. An FTS extends the range of the access network. Transmission delays permitting, an FTS can allow an access network to be located in one country or continent and the host exchange to be located in another.

There is no reason why the functionality of an FTS should not be included in the access network itself. This could be achieved by creating a two-stage access network. The first stage would transport the traffic payloads between the host exchange and various remote locations. The second stage would transport smaller payloads between the remote locations and the final destinations. It would also be possible to carry out the entire transmission in a single step with a very sophisticated optical transmission system, which would have to allow diversity of routing to the remote locations to ensure security and be able to operate with a wide range of distances to the remote ends. If the functionality of the FTS is included in the access network, then there is no FTS functional block in that particular case of the access model, since the FTS has no separate existence. If an FTS is present, then it must be managed as a separate entity, distinct from the access network. If the functionality of an FTS exists, but it is managed as part of the access network, then there is no genuine FTS present, only a more sophisticated access network.

It is also possible to have an rDS between the CPE belonging to an ISDN user and the true access network. In many cases, there is no need for an rDS, because this functionality is included within the access network. The rDS is a legacy of the *digital section* (DS) initially used for ISDN deployment, which consists of an ISDN line termination in the exchange with some form of digital transmission to a remote NT1. A DS is under the control of its associated exchange and is not managed independently of it as part of the access network.

An rDS exists if the line termination of a DS is physically removed into the access network, but the associated control remains in the exchange. In particular, the exchange must be able to activate and deactivate the transmission in the rDS and to monitor the performance of the rDS. If the functionality of an rDS is present, but it is managed as part of the access network, then there is no genuine rDS present, because it is not controlled and monitored by the exchange.

## 9.2 SERVICES AND USER PORTS

It is important to distinguish between the types of user port and the types of services supported at user ports, particularly for ISDN user ports, because physical ports can support different services.

There are four generic service types that can be supported at a user port associated with a V5 interface, although only up to three of these can be supported simultaneously. The first generic type of service is on-demand service, for either ISDN or PSTN, where the connection is set up by the host exchange at the start of each call. In addition to on-demand service, there are two types of leased service. A leased service differs from an on-demand service in that the connection is created by the configuration of the network and not set up for the individual calls.

The first type of leased service is permanently leased service. This is handled by a leased-line network that is separate from the host exchange. User ports that only support permanently leased services are independent of V5 interfaces, because there is no association with a host exchange. The second class of leased service is semipermanent service, where the traffic is routed through a host exchange via a V5 interface, but the connection is set up by network configuration and not for each call. The V5 interface only allows 64-Kbps B-channels to be used for semipermanent services, because narrowband host exchanges are designed to connect 64-Kbps channels. Figure 9.2 summarizes the different generic service types.

User ports that are associated with a V5 interface and that support on-demand services are classified as either PSTN or ISDN user ports, regardless of any leased services supported at the user port. This distinction is only significant for ISDN user ports, because a user port can only be classified as a PSTN port if it supports on-demand PSTN service, and this leaves no channel available for any leased services. The D-channel of an ISDN user port is always connected to the host exchange over the V5 interface because the D-channel contains the call control for the on-demand services. A V5.1 interface can only support basic rate ISDN because it does not have sufficient capacity for standard primary rate ISDN. A V5.2 interface can support both basic rate and primary rate ISDN. Figure 9.3 summarizes the relationship between user ports and services.

User ports that are associated with a V5 interface and that do not support on-demand services are classified as leased ports. These must support either semipermanent service or a combination of semipermanent and permanently leased services, because a leased port that only supports permanently leased service is not connected to an exchange over a V5 interface. Leased ports that only require a single 64-Kbps B-channel on the V5 interface are handled in a similar way to PSTN ports. Leased ports that require more than one B-channel are handled in a similar way to ISDN ports.

| On-Demand | Leased |
|---|---|
| PSTN | Semipermanent |
| ISDN | Permanently Leased |

**Figure 9.2** Generic service types.

| SERVICES | PORTS | | |
|---|---|---|---|
| | PSTN | ISDN | Leased |
| PSTN | Mandatory | n/a | n/a |
| ISDN | n/a | Mandatory | n/a |
| Semipermanent | n/a | Optional | Mandatory |
| Permanently leased | n/a | Optional | Optional |

**Figure 9.3** Services supported at different ports.

The architectural model for services and ports is shown in Figure 9.4. This model ignores the possible FTS, because a FTS is transparent and so has no effect on ports and services. A distinction is made between ISDN ports with and without an rDS, because they have different interfaces at the access network. The nature of the interface to remote NT1s for ISDN is not specified, because it is not internationally standardized. The interface to leased-line equipment is not specified because it also is not standardized.

The host exchange supports the on-demand services (PSTN and ISDN) and the semipermanent leased services. The leased-line network supports the permanently leased services.

## 9.3 V5 LINKS AND TIME SLOT STRUCTURE

The V5 interface can take two forms, V5.1 and V5.2. A V5.1 interface consists of a single 2.048-Mbps link. A V5.2 interface consists of between one and

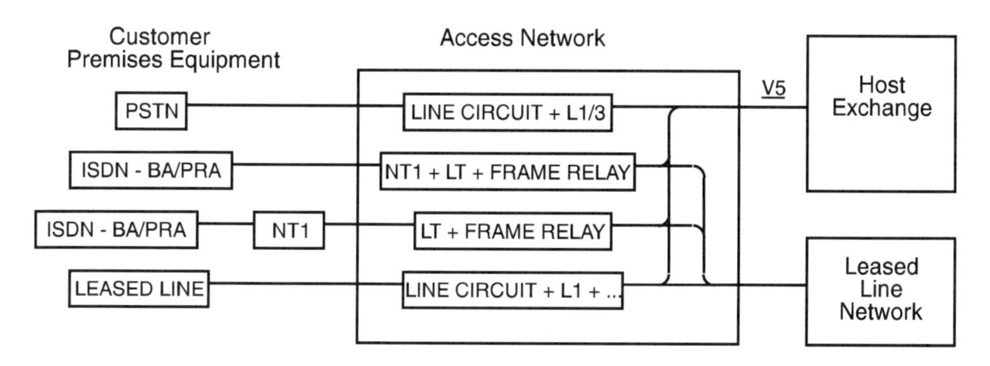


**Figure 9.4** Service architecture.

sixteen 2.048-Mbps links, although in practice a single-link V5.2 interface would be exceptional, because the increased amount of traffic a V5.2 interface supports makes additional links sensible for greater security. In addition to the functions of the V5.1 interface, the V5.2 interface supports both the concentration of traffic and the dynamic assignment of time slots. The 2.048-Mbps links for both interfaces are formatted in the normal way into 32 time slots, with time slot 0 used for frame alignment (see Figure 9.5). A single V5.1 interface can support up to 30 PSTN ports (or 15 ISDN basic access ports), while a single V5.2 interface can support several thousand ports. In both cases, both PSTN and ISDN ports may be supported on the same 2.048-Mbps link.

The V5 interface contains a number of different communications protocols. These are divided into housekeeping communications protocols (control, link control, bearer channel connection, and protection) and call control communications protocols (both for PSTN and ISDN). The call control protocols and the V5 control protocol are relevant to both the V5.1 and the V5.2 interfaces, but the other housekeeping protocols are only relevant to V5.2 interfaces.

ISDN communications are grouped into P-type, F-type, and S-type communications paths. These paths correspond to packet data (SAPI 16), frame data (SAPI 32 to 62), and D-channel signaling (other SAPIs), respectively. Each type of ISDN communication from a single user port is mapped onto a common communications path for that type of information, and each path has an associated V5 communications channel corresponding to a V5 time slot. No two communications paths of the same type can share the same V5 time slot, since communication paths of the same type are only differentiated because they use different time slots. A single ISDN user port always uses the same V5 time slot for each of the three types, but it can use different V5 time slots for different types. Different ISDN user ports may use different communications paths on different V5 time slots for the same type of communication.

Unlike the ISDN communication paths, the housekeeping protocols always share the same V5 time slot. This is time slot 16 of the first 2.048-Mbps link. The PSTN call control protocol also only uses a single time slot, but neither it nor the ISDN communications paths are forced to share the time slot used by the housekeeping protocols to allow extra bandwidth to be allocated to

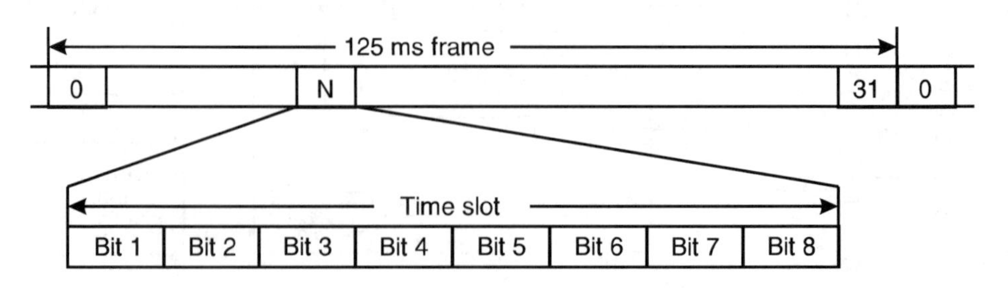


**Figure 9.5** Format of 2.048-Mbps links at V5 interfaces.

call control as the number of user ports increases or as ISDN D-channel traffic increases.

### 9.3.1 The V5.1 Interface

For V5.1, there is only a single S-ISDN (signaling) communications path with a corresponding unique V5 time slot, which may or may not be shared by other communications protocols or different types of ISDN communications paths. There may be a number of different P-ISDN (packet data) and F-ISDN (frame relay) communications paths making use of up to three time slots. If there is only one time slot used for all communications channels, then it must be time slot 16, because the control protocol is located there.

If two time slots are used for communications, then they must be time slots 16 and 15. The control protocol must use time slot 16 and at least one of the other communications paths must obviously use time slot 15 (see Figure 9.6). For instance, there may be F-ISDN communications paths on both time slots. Likewise, there may be P-ISDN communications paths on both time slots. The PSTN protocol and the S-ISDN communications path may each use either time slot 16 or 15.

If three time slots are used for communications, then these are time slots 16, 15, and 31 (see Figure 9.7). Once again the control protocol must use time slot 16. Since the PSTN protocol can only use a single time slot, there must also be ISDN communications present if three time slots are used. There may be F-ISDN and P-ISDN communication paths on any of the time slots. The PSTN protocol and the S-ISDN communications path may each use either time slot 16, 15, or 31.

### 9.3.2 The V5.2 Interface

In addition to the difference in the number of 2.048-Mbps links, the V5.2 interface differs from the V5.1 interface in two major ways. First, the V5.2

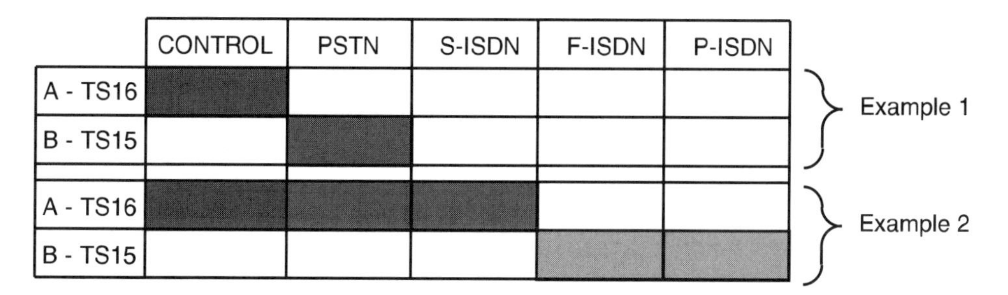


**Figure 9.6** Possible assignments for V5.1 with two communications time slots.

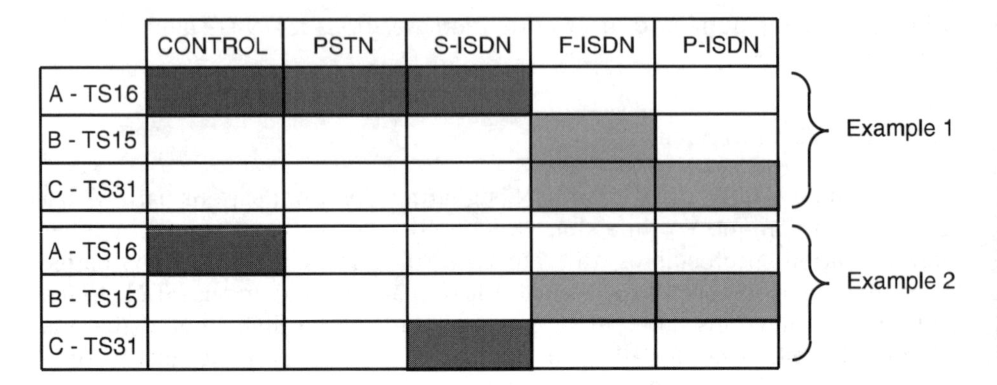


**Figure 9.7** Possible assignments for V5.1 with three communications time slots.

interface supports additional housekeeping protocols, which share the same time slot as the control protocol. The second major way in which the V5.2 interface differs from V5.1 interface is that it has additional backup time slots to improve the security of the communications (see Figure 9.8). Apart from these major differences, V5.2 differs from V5.1 in that there may be more than one S-ISDN communications path, so that ISDN call control is not constrained to a single V5 time slot. This is necessary to allow extra bandwidth to be allocated to call control for the additional ISDN ports the V5.2 interface can support. Extra bandwidth for the additional PSTN ports was not thought to be necessary, since PSTN signaling should be less demanding on bandwidth.

The introduction of the additional housekeeping protocols may have an indirect impact on the allocation of communications paths to time slots, because they reduce the spare capacity on the time slot used by the control protocol. Because of the presence of the additional protocols, call control

| Timeslots | Primary Link | Secondary Link | Other Links |
|---|---|---|---|
| 15 | (optional) | (optional) | (optional) |
| 16 | Housekeeping Protocols | Housekeeping Protection | (optional) |
| 31 | (optional) | (optional) | (optional) |

**Figure 9.8** V5.2 communications time slots.

communications are less likely to also share the same time slot, particularly if they are heavily used. There is no direct impact on the allocation of protocols to time slots due to the introduction of the additional housekeeping protocols, because all housekeeping protocols are effectively allocated to a single communications time slot as a single compound protocol.

The use of backup channels has an effect similar to the allocation of additional time slots to a communications path. The net effect is to increase the number of time slots associated with the communications. The only change to principles involved in the V5.1 case is that backup makes it necessary for a communications path to be dynamically associated with more than one time slot, preferably on different links in case there is a failure on a specific link.

#### *9.3.2.1 Protection of Communications Channels*

The V5.2 interface has the ability to automatically protect the logical communications channels that carry signaling and housekeeping protocols between the access network and the host exchange. This feature allows the V5.2 interface to survive the failure of one of its component links, because the communications on the failed link can be automatically switched to another link. This assumes, of course, that a V5.2 interface consists of at least two links.

Protection is performed for specified 64-Kbps logical communications channels and includes all communications paths assigned to that channel. Protected communications channels belong to either protection group 1 or to protection group 2.

Protection group 1 handles the logical communications channel containing the housekeeping protocols and uses time slot 16 on both the primary and the secondary V5.2 links (see Figure 9.8). This logical communication channel is the main V5.2 communications channel and the two time slots are the physical communications time slots with which it can be associated. Initially the main communications channel is associated with time slot 16 on the primary V5.2 link.

The V5.2 protection protocol monitors time slot 16 on both the primary and the secondary V5.2 links. This ensures that degradation on the primary link is detected and that the availability of the secondary link is ensured. If the performance on the primary link drops too far, then the main logical communication channel is switched to time slot 16 of the secondary link. It is possible that a few messages will be corrupted during this switch-over, but this is not significant, because the corruption will be detected and the messages retransmitted.

Logical communications channels other than the main channel can be protected by including them in protection group 2. Protection group 2 differs from protection group 1 in that protection group 2 does not have a standby time slot for each active time slot and the V5.2 protection protocol is not transmitted on the backup time slots. There are no more than three standby time slots in protection group 2, because three are sufficient to provide protection in the event of a single-link failure. Any number of active communications time slots, which are not otherwise protected, may be assigned to protection group 2. Logical communications channels in protection group 2 will be switched to a free standby time slot in the event of a failure of their original time slots. As for protection group 1, any messages that are corrupted as a result of the switch-over will be retransmitted as part of the normal error correction process.

The protection protocol coordinates the switch-over for both protection groups, so that both sides of the interface switch in the same way. Logical communications channels other than the main communications channel can be left unprotected.

#### *9.3.2.2 Control of the 2.048-Mbps Links*

The multiple links of a V5.2 interface are managed across the V5.2 interface by its link control protocol. This protocol allows links to be identified and blocked or unblocked.

Link identification is required to verify the integrity of the physical connections of the V5.2 interface. It operates by tagging the link to be identified, in the same way that a tone may be applied to a specific copper pair in cable. In this case a digital signal is applied instead of a tone.

Link blocking and unblocking are required to allow links to be maintained with the minimum disruption of traffic and to allow the interface to evolve as the traffic evolves. This is almost identical to the blocking and unblocking of user ports on the control protocol.

## 9.4 BEARER TIME SLOTS AND V5 TRAFFIC CAPACITY

The bearer time slots on a V5 interface are used to carry 64-Kbps circuit-switched traffic from the user ports to the host exchange. These time slots must be allocated to the user ports in a way that is clearly agreed-on, so that both the access network and the host exchange know which time slots are used for a particular user port.

For a V5.1 interface, the allocation of bearer time slots to user ports is static, but may be reconfigured over the management interfaces at the access network and at the host exchange. Here the term *static* is used in the sense that the allocation does not change from call to call. There is a one-to-one mapping between the appropriate bearer channels at the user ports and the bearer time slots on V5.1 interfaces. For a simple-access network, this mapping may be hard-wired.

For a V5.2 interface, the allocation of bearer time slots to user ports is dynamic and will normally change from call to call. The mapping between the

V5-related bearer channels at the user ports and the bearer time slots on the V5.2 interface is controlled by the V5.2 BCC protocol. Bearer time slots are allocated to user ports flexibly according to demand. This flexibility gives greater security and supports the concentration of traffic.

The dynamic allocation of bearer time slots on a V5.2 interface gives greater security, because service is maintained even if a link is lost. This requires, of course, that there is more that one link on the V5.2 interface. Individual calls may be lost if a V5.2 link fails, but these can be reestablished on a different link if the user redials. The quality of service after a failure will be less because the traffic is supported by fewer time slots. This increased security is impossible on a V5.1 interface, because its static allocation ties the service to the lost bearer time slots.

The dynamic allocation of bearer time slots on a V5.2 interface also supports the concentration of bearer traffic. The interface can support more bearer channels at the user ports than bearer time slots on the V5.2 interface. Concentration takes advantage of the fact that only a fraction of all user ports are likely to be active at any given time. For reasonably large systems, a concentration factor of about 8 is routinely applied, because it creates no perceived reduction in the quality of service. This allows an access network with about 1,000 PSTN ports to be supported by a single V5.2 interface with only four links.

A single V5.2 interface is capable of supporting about 4,000 PSTN ports, because it may have 16 links and concentrate the bearer traffic by a factor of 8. A single V5.1 interface is only capable of supporting 30 PSTN ports, because at least one time slot on the link is required for signaling and another is required for frame alignment.

Dynamic allocation of V5 bearer time slots is not identical to the concentration of bearer traffic, because dynamic allocation does not determine the ratio of user ports to bearer time slots on the V5 interface. In theory, dynamic allocation can be used with more time slots than those required if all user ports were busy, but this is unrealistic, except perhaps when a V5 interface is newly installed and only a few customers have been allocated to it. It is not necessary for a V5.2 interface to be concentrating, but it must have dynamic allocation of bearer time slots.

Concentration of traffic across the V5.2 interface is different from concentration of traffic within the access network itself. The transmission system of the access network need not concentrate traffic, even if the V5.2 interface is concentrating. A concentrating V5.2 interface may be used regardless of the details of the transmission within the access network, because the cost per user port of the interface can be less despite the greater complexity of the interface, since fewer links are required to support the traffic. Even if the access network is concentrating, the concentration can be hidden within the access network, which could use a nonconcentrating V5.1 interface. For example, an access network using radio transmission may use concentration on the transmission because radio bandwidth is limited, but use a nonconcentrating V5.1 if the size of the system is too low to justify the added complexity of a V5.2 interface.

## 9.5 SUMMARY

The V5 interface is the interface between an access network and a host exchange for the support of narrowband telecommunications services. It has two forms, V5.1 and V5.2, and the more sophisticated V5.2 form supports the concentration of traffic. The architectural model of the access network used in the definition of V5 interfaces allows an access network to be linked to its host exchange by an FTS and it allows ISDN ports to be connected to the access network by an rDS. Although both FTSs and rDSs are permitted, neither is required.

Both V5.1 and V5.2 interfaces support both PSTN and ISDN user ports, but a V5.2 interface is required if primary rate ISDN is to be supported. V5 interfaces also support leased-line ports with semipermanent leased lines connected via the exchange by treating them like PSTN or ISDN lines. The V5 interface also allows an access network to have ISDN ports with certain B-channels nailed up for leased service. The traffic on these nailed-up B-channels may be routed through a service interface that is not a V5 interface, for instance, to a leased-line network.

V5.1 interfaces consist of a single 2.048-Mbps link, while V5.2 interfaces may consist of up to 16 such links. Both types of V5 interfaces may use time slots 15, 16, and 31 for signaling, but there are detailed constraints on the allocation of signaling to these time slots.

V5.2 interfaces are able to concentrate bearer traffic so that the bearers at the user ports are supported by a smaller number of bearer channels on the interface. The allocation of bearer channels to user ports on a V5.2 interface may change with each call, unlike the allocation for a V5.1 interface. This dynamic allocation of bearer channels also makes the V5.2 interfaces more secure. It does not determine the amount of concentration at the interface, and whether or not concentration is used at the V5.2 interface is not determined by whether concentration is or is not used within an access network.

V5.2 interfaces also allow their individual links to be identified so that the integrity of the interface can be checked. Individual links can also be removed from service, either temporarily for maintenance or permanently to make the interface smaller. Likewise, new links for a V5.2 interface can be brought into service.

The V5.2 interface provides for protection of the time slots reserved for signaling. The time slot used for the most sensitive protocols is always protected—the active and standby time slots forming protection group 1. Other signaling time slots may also be protected, and their active and standby time slots form protection group 2.

## Selected Bibliography

*V5.1 Interface Specification for the Support of Access Networks*, ITU-T Recommendation G.964.

*V5.1 Interface Specification for the Support of Access Networks*, ETSI Specification ETS 300 324-1.

*V5.2 Interface Specification for the Support of Access Networks*, ITU-T Recommendation G.965.

*V5.2 Interface Specification for the Support of Access Networks*, ETSI Specification ETS 300 347-1.

# V5 Multiplexing and Message Formats

# 10

"They sought it with thimbles, they sought it with care."
—Lewis Carroll, *The Hunting of the Snark*

The previous chapter on architecture and interfaces introduced the link structure and the protocols of the V5 interface. The structure of the V5 messages and how these are multiplexed onto time slots of the V5 interface are now examined in more detail, based on the general principles involved and how these are applied in the case of the V5 interface. The multiplexing of ISDN call control messages is described in more detail in the next chapter, followed by the detailed examination of the various protocols used on V5 interfaces.

The physical multiplexing, corresponding to *Open Systems Interconnection* (OSI) layer 1, is discussed first. The addressing concepts, which involve both the frame layer (OSI layer 2) and the message layer (OSI layer 3) are then explained. The common message structure of the V5 protocols is described next, followed by the explanation of the message labeling convention used in subsequent chapters.

## 10.1 PHYSICAL MULTIPLEXING

The physical multiplexing used on V5 interfaces has two different aspects. It includes both the physical structure of the links on the interface, and how the physical structure of the links is used to support the different communications protocols.

It is also necessary to distinguish between the logical communications channels and the physical communications time slots to which they are mapped, because this mapping can be altered by protection switching on V5.2 interfaces.

### 10.1.1 Structure of the Links

There are a number of standard physical links that are used for multiplexed communications in telecommunications systems. The most common physical link in the United States and Japan operates at 1.544 Mbps, while in Europe and much of the rest of the world the equivalent link operates at 2.048 Mbps. The V5 interface was specified to consist of 2.048 Mbps links because it originated in Europe. There is no technical reason why the V5 interface cannot be at the 1.544-Mbps rate, because both rates are based on 64-Kbps time slots, although the difference in the details must also be specified. The possibility of higher rates for the physical links, such as SDH rates above 2.048 Mbps, was considered but was thought to be unnecessary for the first generation of systems.

The 2.048-Mbps links are formatted into 32 time slots (see Figure 10.1). This formatting is achieved by using time slot 0 as a marker, which leaves each of the remaining 31 time slots free to carry a 64-Kbps physical channel. The interfaces use these physical channels to carry bearer traffic, which is routed end to end across the telecommunications network, and the various communications protocols used between the access network and its host exchange. The simplest way to achieve this is to identify 64-Kbps time slots for use either as a bearer time slot or as a communications time slot.

It is natural to use time slot 16 of a 2.048-Mbps link as a communications time slot, because time slot 16 is used in this way in other standards. If several communications channels are required, then time slot 16 in a number of 2.048-Mbps links can be used. Using time slot 16 on different links for communications channels might not provide sufficient bandwidth, because ISDN users could make heavy use of their D-channel communications protocol and because the interface may consist of only a small number of links. The need for additional bandwidth for signaling is greater if the interface is capable of concentrating user traffic so that it can support a large number of user ports.

The guiding principle for the allocation of physical time slots for use by communications protocols is to minimize the number of links with an odd number of bearer channels, because this simplifies the mapping of ISDN ports onto the links. It is also sensible to allocate time slots 15 and 31 of a single link

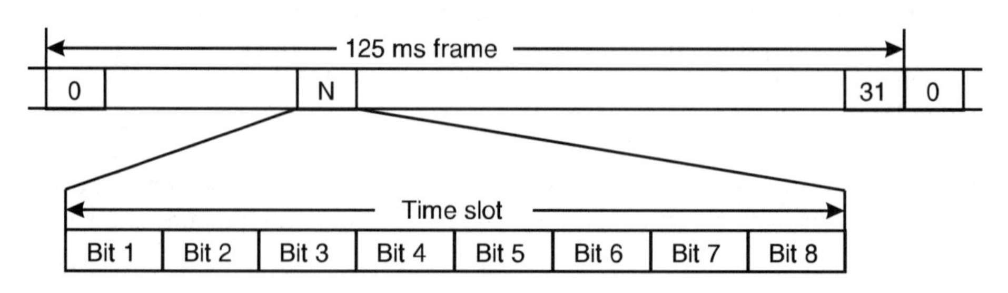


**Figure 10.1** Format of the 2.048-Mbps links for V5.

if time slots other than time slot 16 are required, because this is compatible with the specification of ISDN primary rate ports when time slots are assigned in blocks.

Time slot 16 of each link in turn should be used first, because this creates an even number of bearer time slots by balancing the use of time slot 0 for alignment. Once all time slots 16 are allocated for use as bearer channels, the number of links with odd numbers of bearer channels is minimized by allocating time slot 15 and 31 of each link in turn (see Figure 10.2). The V5 specification recommends this allocation, but does not insist on it, because rigorous adherence could cause problems as the interface evolves or when protection switching of communication channels is performed.

### 10.1.2 Use of Physical Communications Time Slots

The communications protocols between an access network and its host exchange are either used for user signaling or for housekeeping functions. For the V5 interface, the control, BCC, link control, and protection protocols are the defined housekeeping protocols. The PSTN call control protocol has also been defined for PSTN user signaling, because unlike the situation for ISDN, no existing standard could meet all of the requirements. All V5 protocols need to be assigned to communications channels.

The number of communications time slots needed is determined by the bandwidth required by the communications protocols. Additional physical communications time slots are also required to support the logical communications channels when there is protection switching. The allocation of communications time slots for the ISDN call control protocol is described in more detail later because it is necessary to take account of the generation and reception of messages in ISDN terminals that are outside of the access network, whereas all the other protocols terminate within the access network. The allocation of the communications channels for these other protocols is described in this chapter.

Ideally, no more than one time slot should be used for a communications protocol, other than for ISDN signaling, because otherwise there is a need to coordinate the messages on different time slots. This is not a problem for ISDN signaling because frame relaying allows these messages to be distinguished by their frame layer addresses. The protection protocol is an exception to this rule, because it is desirable for the protection protocol to be assigned to more than one time slot so that it can continue to operate if one of its time slots fails. More than one protocol can be assigned to a channel or time slot because different protocols can be distinguished at frame and message levels.

If bandwidth permits, then all housekeeping protocols should be assigned to the same logical communications channel, because this makes protection of these protocols simpler. Call control protocols need not share this logical

| Primary Link | | | Secondary Link | | | Link 3 | | | Link 4 | | |
|---|---|---|---|---|---|---|---|---|---|---|---|
| 15 | 16 | 31 | 15 | 16 | 31 | 15 | 16 | 31 | 15 | 16 | 31 |
| | X | | | | | | | | | | |
| | X | | | X | | | | | | | |
| | X | | | X | | | X | | | | |
| | X | | | X | | | X | | | X | |
| X | X | | | X | | | X | | | X | |
| X | X | X | | X | | | X | | | X | |
| X | X | X | X | X | | | X | | | X | |
| X | X | X | X | X | X | | X | | | X | |
| X | X | X | X | X | X | X | X | | | X | |
| X | X | X | X | X | X | X | X | X | | X | |
| X | X | X | X | X | X | X | X | X | X | X | |
| X | X | X | X | X | X | X | X | X | X | X | X |

**Figure 10.2** Sequence of using communications time slots.

channel and may have a lower level of protection because a failure of a single call control protocol will not necessarily affect all user ports.

The housekeeping protocols of the V5 interface are initially assigned to time slot 16 of its primary 2.048-Mbps link. For V5.1 interfaces, this is the only 2.048-Mbps link and the only housekeeping protocol is the V5 Control protocol. For V5.2 interfaces, time slot 16 of the primary link also contains the BCC protocol and the link control protocol, because initial estimates suggested that a single time slot has sufficient bandwidth. The V5.2 protection protocol is assigned to time slot 16 of both the primary and the secondary 2.048-Mbps links to ensure that the protection protocol is automatically immune to the failure of a single link.

The PSTN call control protocol can be assigned to any physical communications time slot that is not reserved for protection. This protocol can be assigned to the same logical communications channel used for the general control protocols if sufficient bandwidth remains available. Alternatively, the entire bandwidth of a communications channel may need to be dedicated to the PSTN protocol if the protocol is heavily used. This flexibility allows the interface to be tailored to different applications. Similar considerations also apply to ISDN signaling, but the details differ because of the nature of ISDN signaling.

## 10.2 FRAME AND MESSAGE LEVEL MULTIPLEXING

Messages can be multiplexed onto different 64-Kbps communications channels, as described above. This type of multiplexing is at the physical layer (OSI layer 1). Within a communications time slot, messages can be multiplexed at the frame layer (OSI layer 2) and at the message layer (OSI layer 3).

One of the purposes of the frame layer is to adapt the specific characteristics of the physical layer to provide error-free communications, which are independent of the physical medium, to the message layer and to allow a number of message layers to be supported by the same physical layer. This is achieved by addressing and numbering the messages and by adding a CRC tail as a checksum so that errors can be detected. When an error is detected, a request is sent to repeat from the last correctly received message.

The V5 interface makes special use of the address space at the frame and message layers.

### 10.2.1 Addressing Requirements

A frame relay approach to the multiplexing of ISDN call control protocols has been adopted because it is not necessary for the access network to interpret the content of ISDN call control messages. PSTN messages, on the other hand, must be interpreted by the access network, because they need to be mapped onto signals which are intelligible at a PSTN user port (e.g., off-hook and ringing). This means that the frame layer for ISDN messages terminates in the ISDN terminals, whereas the frame layer for PSTN messages terminates in the access network. For call control messages, ISDN ports are identified by the frame layer addressing, whereas PSTN ports must be identified by message layer addressing.

On the other hand, it is also sensible to be able to refer to ports independently of whether call control uses a frame layer or a message layer address. This is particularly important for any control protocol messages that need to refer to either ISDN or PSTN ports.

### 10.2.2 The General V5 Address Space

The addresses used on the V5 interface at frame and message layers have been chosen to allow the control protocol to refer to both ISDN and PSTN user ports using message layer addresses that are the same as the addresses used for call control at either frame or message layers. This approach creates relationships between frame and message layer addresses. The result is a general address space that is mapped onto frame and message address spaces for different messages (see Figure 10.3).

The general V5 address space contains addresses for ISDN ports, PSTN ports, and each of the V5 protocols. The general space addresses for ISDN ports correspond to the frame layer addresses used to differentiate between ISDN ports. The message layer addresses for ISDN are defined in the standard specifications for the ISDN protocol and are outside the scope of the V5 specification. The general space addresses for PSTN correspond to the message layer addresses used to differentiate between PSTN user ports for PSTN call control. The general space address for the PSTN protocol, which is different from the general space addresses for the PSTN ports, is the frame layer address used for PSTN call control messages.

Control protocol messages make use of the general space addresses for PSTN ports, ISDN ports, and the control protocol itself. These addresses are used for the message layer addressing of the control protocol messages. The message layer address for control protocol messages is used to indicate whether the control message relates to a PSTN port or an ISDN port, or if it is a general control message. The frame layer address used for control protocol messages is the general space address for the control protocol.

The general space address for the control protocol has a dual function. When it is used at the frame layer, it indicates that the message belongs to the control protocol. When it is used at the message layer, it indicates that the control message is not associated with a user port, but is a general control message.

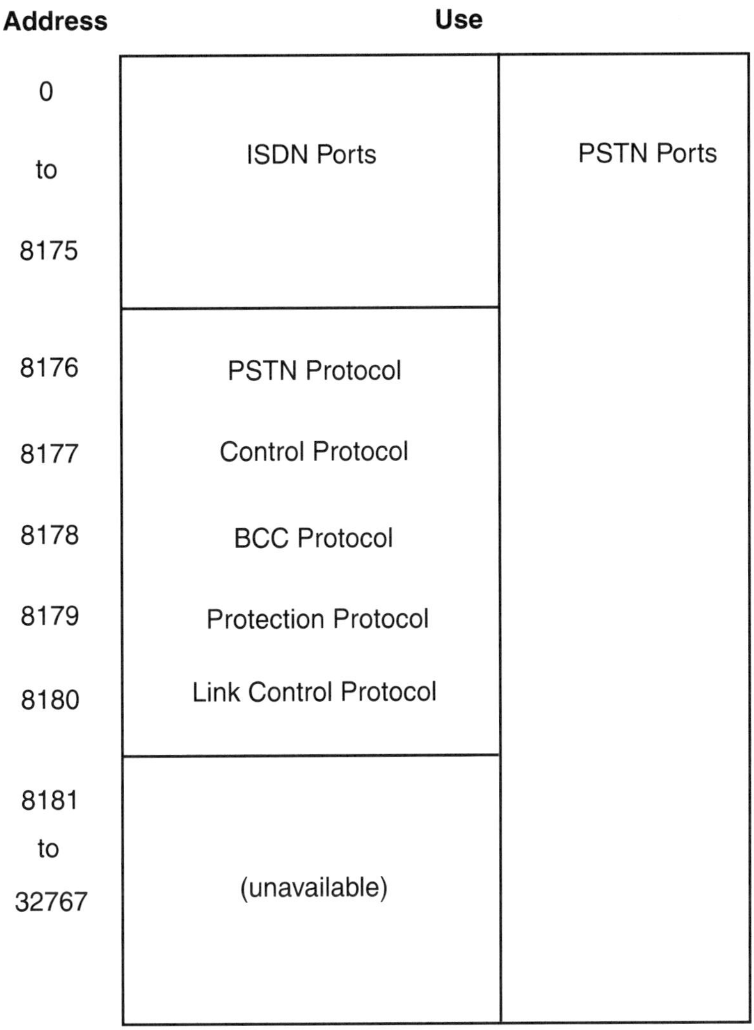

| Address | Use | |
|---|---|---|
| 0 to 8175 | ISDN Ports | PSTN Ports |
| 8176 | PSTN Protocol | |
| 8177 | Control Protocol | |
| 8178 | BCC Protocol | |
| 8179 | Protection Protocol | |
| 8180 | Link Control Protocol | |
| 8181 to 32767 | (unavailable) | |

**Figure 10.3** General frame and message layer address space.

The general space addresses for the V5.2-specific protocols are only used at the frame layer. They are used to indicate that the messages are not associated with user ports, but belong to the identified V5.2-specific protocol. The format of the associated message layer addresses for the V5.2-specific protocol, like the message layer addresses for ISDN signaling messages, is not related to the general address space.

## 10.3 FRAME LAYER ENVELOPES

A two-level structure has been specified for the V5 frame layer so that ISDN call control messages can be frame-relayed across the access network. The frame layer needs two levels, because ISDN messages already have a frame layer, and so a second outer part of the frame layer must be introduced as an envelope with a label to indicate the correct ISDN user port for the contained ISDN signaling. For ISDN call control messages, the labeling address for this envelope is the ISDN port address from the general V5 address space.

The same two-level frame addressing structure which has been specified because ISDN messages already have a frame layer has also been adopted for the messages of the V5 protocols. For messages other than ISDN call control, it is necessary in addition to specify the inner part of the frame structure, because it is not already specified, since the other protocols are new.

The general form of a V5 frame envelope is shown in Figure 10.4. The outer envelope address is a 13-bit number that, with three fixed bits, makes up the two octets that follow the opening flag of the frame. This allows the envelope address to take the values 0 to 8191, which is consistent with that part of the general V5 address space. The remaining bits in octets 2 and 3 of the frame are the *extension address* (EA) bits and the bit corresponding to the *command/response* (C/R) bit of the ISDN frame, but which is fixed at 0 here since its function is performed by the C/R bit of the contained inner data link sublayer.

Envelope addresses in the range 0 to 8175 are used to identify the ISDN ports associated with the V5 interface. The remaining envelope addresses, 8176 to 8191, are used to identify the virtual ports within the equipment at either side of the V5 interface where the V5 protocols terminate, according to the general V5 address space. The V5 frames end with two frame check octets, followed by a closing flag. There is no frame check sequence for the inner data link sublayer (see Figure 10.5), since there is no need for the frame integrity to be checked twice.

The contained information field of the data link sublayer lies between the outer address of the frame envelope and the start of its frame check octets. For ISDN call control messages, this information field begins with the frame layer address fields of an ISDN call control message. For other protocols, this also begins with two octets containing an address, the data link address. These octets also contain the EA bits and the C/R bit used in the same way as for ISDN frames. These two octets are followed by the control octets, which may be followed by a data link information field.

Like the outer envelope address, the inner data link address for protocols other than ISDN call control also consists of 13 bits, which allow it to take the range 0 to 8191. The envelope address and the data link address contain the same information for these protocols. Addresses in the range 8176 to 8180 indicate the PSTN protocol, the control protocol, the BCC protocol, the protec-

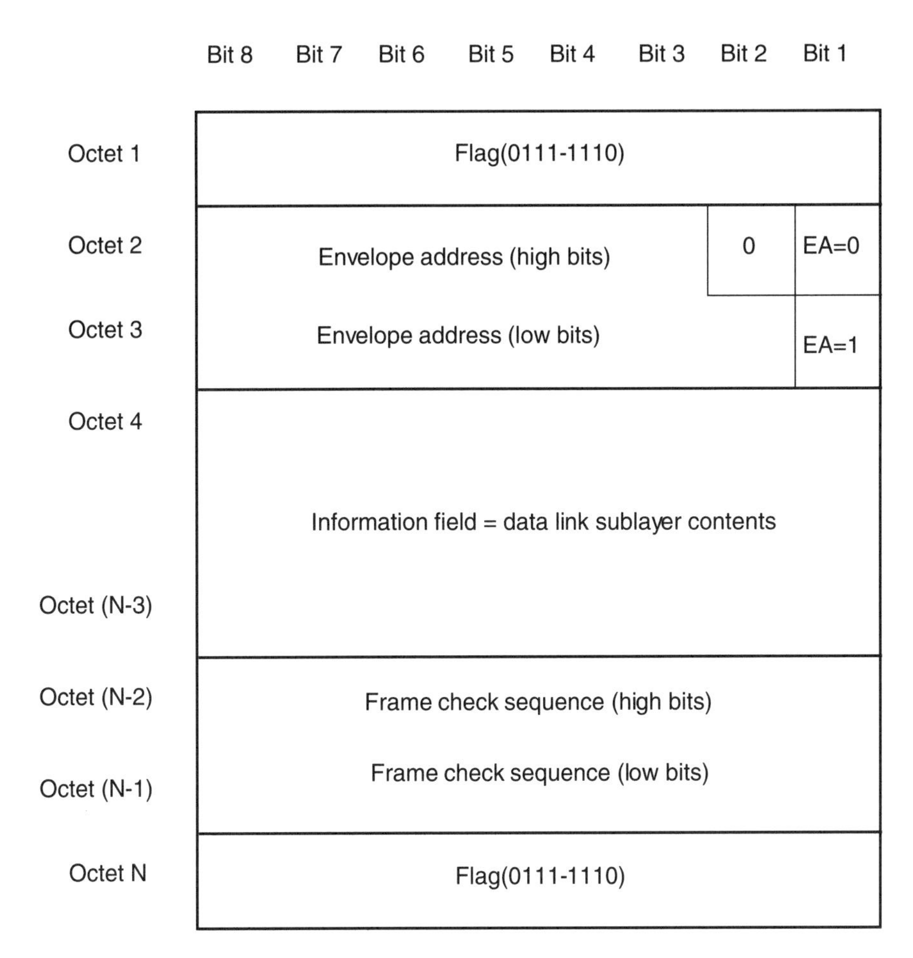


**Figure 10.4** Outer frame envelope for V5.

tion protocol, and the link control protocol, respectively, as defined in the general V5 address space.

If the frame layer addresses identify a V5 protocol, then the data link information field contains the message layer information of the V5 protocol.

## 10.4 LAYER 3 MESSAGE FORMATS

Within the data link information field, all V5 protocols use a similar format at the message layer (see Figure 10.6). The message format for ISDN signaling

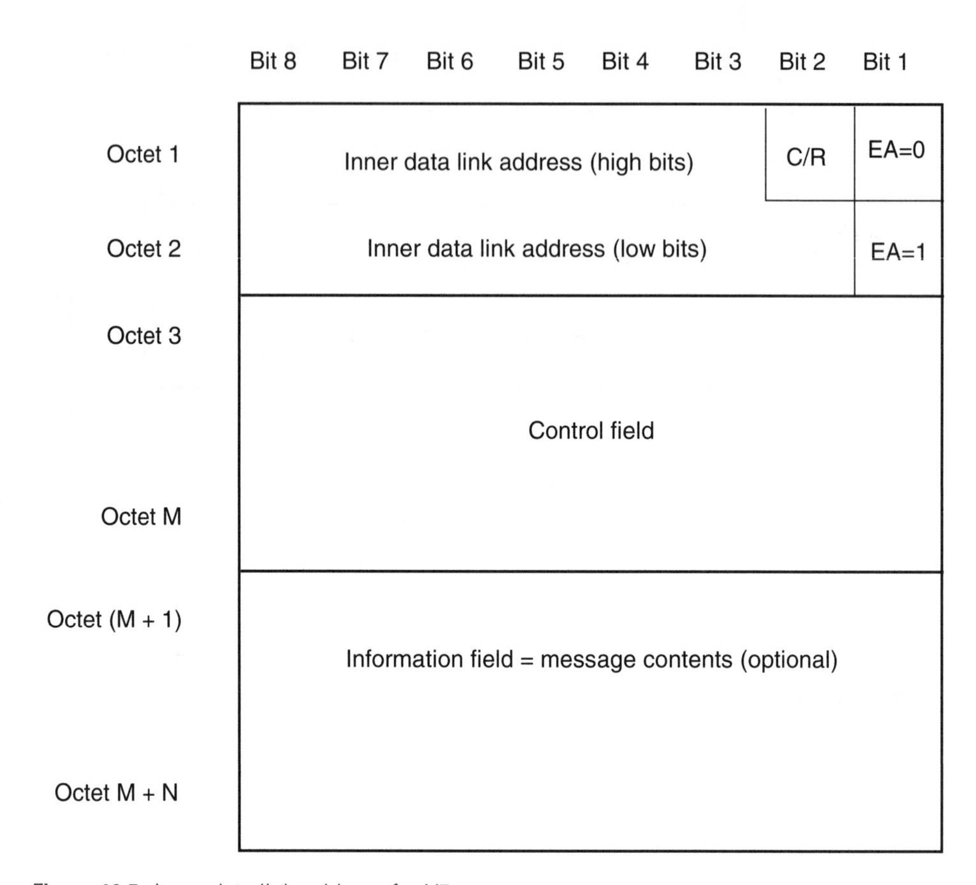


**Figure 10.5** Inner data link sublayer for V5.

messages is different because the messages for ISDN were defined independently and their message layer information is conveyed transparently in a V5 frame layer envelope.

All V5 protocol messages begin with mandatory Protocol and Address information elements. These are followed by the mandatory Type information element. Following these mandatory information elements there may be additional information elements, but which information elements are included, and under what conditions they are included, are determined by the message type and by which side of the interface is sending the message.

The first octet of a V5 protocol message is the V5 protocol discriminator, which has the value 72 (i.e., 48 in hex notation). This is used to prevent a V5 protocol message from being confused with messages in any other protocols that may be specified in the future. The fourth and final octet of the message

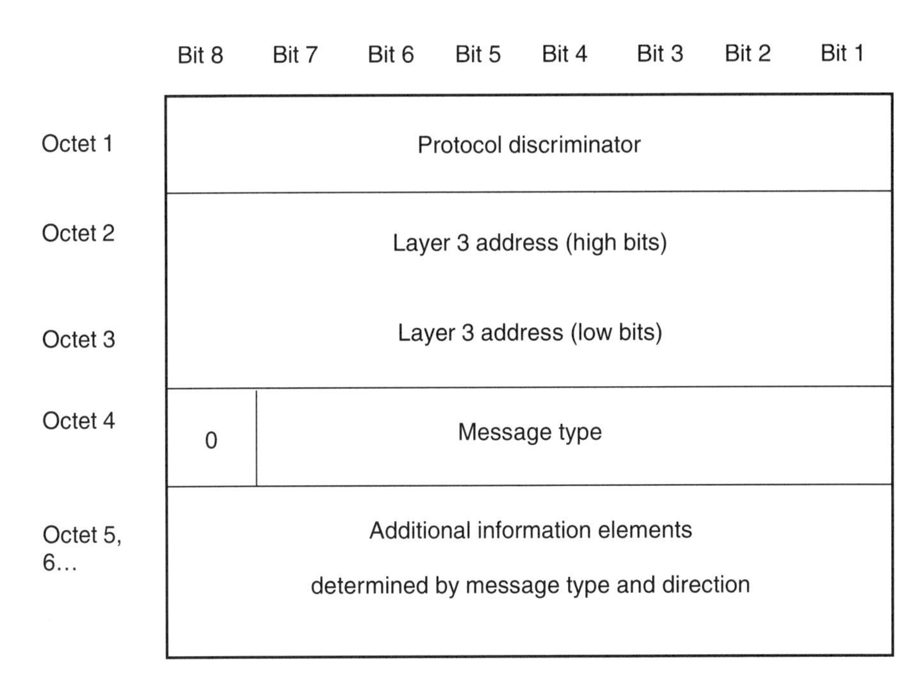


**Figure 10.6** V5 protocol message format.

header is the message type, which uniquely identifies the different V5 messages and so implicitly identifies the V5 protocol to which they belong. The protocol to which the messages belong is also identified explicitly in the frame layer addresses.

### 10.4.1 Message Layer Addresses

Between the protocol discriminator and the message type are the two octets that contain the message layer address. The meaning of the message layer address depends on the V5 protocol identified by the frame layer address.

For the control protocol, the message layer address takes a value from the general V5 address space. A bit in this address indicates whether the remaining bits are a PSTN user port address for port-related control messages. Otherwise, they form an ISDN user address for ISDN port-related control messages, or they indicate the control protocol itself for common control messages (see Figure 10.7). For the PSTN protocol, the message layer address takes a value from the general V5 address space that identifies a PSTN user port. The two octets of the field contain 15 usable bits and one fixed bit. All possible combinations of these

| Protocol | Message address |
|---|---|
| PSTN | PSTN port address |
| Control (port)<br>Control (common) | ISDN or PSTN port address<br>Control protocol address |
| BCC | Process reference number and source |
| Protection (except SN messages)<br>Protection (SN messages) | Logical communications channel identifer<br>(all zeros) |
| Link control | Link identifier |

**Figure 10.7** V5 message layer addressing.

15 bits are allowed, giving a V5 interface the potential to carry the signaling for up to 32,768 PSTN ports.

For the BCC protocol, the message layer address identifies a BCC process. The two octets contain 13 usable bits plus one bit to indicate either the access network or the host exchange. The 8,192 possible values of the 13 usable bits indicate the BCC process to which the message refers. For the link control protocol, the message layer address contains only eight usable bits. These bits take the value of the link identifier used for the BCC protocol to specify one of the 16 V5.2 links. For the protection protocol, the message layer address can make use of all 16 bits of the two octets. The value used identifies the logical communications channel to which the message refers.

### 10.4.2 Message Types

There is a relationship between the V5 addresses used at the frame layer and the message type information element in the generic header. The addresses used at the frame layer and the fields used for message types differ because there is a one-to-many mapping (see Figure 10.8). There are many values of the

| Protocol | Frame address | Message type coding |
|---|---|---|
| PSTN | 8176 | 0 to 15 |
| Control | 8177 | 16 to 23 |
| BCC | 8178 | 24 to 31 |
| Protection | 8179 | 32 to 47 |
| Link control | 8180 | 48 to 55 |

**Figure 10.8** V5 frame addresses and message types.

message type field for each value of the frame address because there are many messages for each protocol specified at the frame layer.

The values of the message layer fields for the different V5 protocols do not overlap, and this is possible because there are fewer message types in total than possible message type values. This means, however, that there is redundant information, because it is possible to deduce the frame layer address from the message type. This also allows the possibility of an inconsistency occurring between a frame layer address and a message type field.

## 10.5 CONVENTIONS USED FOR IDENTIFYING MESSAGES

The following convention will be used later in the analysis of the various V5 protocols, where it is sometimes necessary to describe messages at various levels of detail. Sometimes it is sufficient to simply refer to the type of message, but at other times it is necessary to also indicate information elements that are present. It is even necessary at times to refer to specific values of the information elements.

Messages are identified principally by their type, which is written in capitals and hyphenated if there is more than one word. For example, there is an ESTABLISH message in the PSTN protocol and a PROTOCOL-ERROR message in the protection protocol. If it is necessary to specify which side of the interface sends the message, then this is denoted by adding a prefix. For example, an AN/ESTABLISH message is sent by the access network and an LE/ESTABLISH message is sent by the exchange. This may be useful because the message-specific information elements that are permitted can depend on which side of the interface sends the message.

Optional message-specific information elements are indicated by adding a suffix that begins with a capital and is hyphenated if there is more than one word. For example, ESTABLISH/Steady-signal is an ESTABLISH message containing a Steady-signal information element. If there are optional information elements, but none are included, then this is indicated by a dash; that is, AN/ESTABLISH/- is an ESTABLISH message sent by the access network that contains no optional information elements.

The values of optional information elements are indicated by extending the suffix with a colon and its value. For example, LE/ESTABLISH/Steady-signal:cadenced-ringing is an ESTABLISH message sent by the exchange that contains the optional Steady-state information element, and this optional information element has the value that represents "cadenced-ringing."

The values of mandatory information elements may also be indicated in the same way as optional information elements. In addition, the notation may be abbreviated, because it is not necessary to specify that a mandatory information element is present. Thus, STATUS:Response:AN0 is a STATUS message with a Cause information element that indicates that it has been sent in response to an LE/STATUS-ENQUIRY message and that indicates that the PSTN port in the access network (specified by the message layer address in the generic header) is in state 0 (out of service).

Abbreviation may also be used for optional information elements. In this case, the value implies that the optional field is included. Thus, ESTABLISH/Line-information:impedance-marker-set is equivalent to ESTABLISH:impedance-marker-set because the optional Line-information element must be present by implication because its value is given.

It should be noted that this convention allows messages that are not valid for the V5 interface to be identified. For example, an LE/STATUS message is not a valid V5 message because the exchange is not allowed to send a STATUS message. If only valid V5 messages are involved, then PROTOCOL-PARAMETER and LE/PROTOCOL-PARAMETER are equivalent, because an AN/PROTOCOL-PARAMETER message would violate the V5 interface specification.

The specific V5 protocol to which messages belong is not indicated in this convention, because these protocols are identified by the frame layer address. However, the protocol is identified by implication through the message name. This is consistent with the approach taken for the V5 interface, since the message type information element in the generic header, which contains the coded value of the message name, also implicitly identifies the protocol that is defined explicitly by the frame layer address.

## 10.6 GENERAL OBSERVATIONS

The physical layer multiplexing on the V5 interface is a natural consequence of using 2.048-Mbps links, since the use of the time slots is compatible with earlier telecommunications standards. The approach is also capable of being extended to other physical layers, such as 1.544-Mbps links and SDH links, so long as the other links are based on the multiplexing of 64-Kbps channels.

The frame relay approach to the handling of ISDN call control signaling is effective because it avoids unnecessary interpretation of this signaling within the access network. This frame relay approach could be extended later to other types of signaling that need not be processed within an access network. The V5 approach to frame relaying assumes that there are CRC octets at the end of the frame that is relayed, and uses these to avoid relaying frames that have been corrupted. This may need to be modified if different information, for example, containing forward error correction coding, were to be frame relayed.

The frame level multiplexing of the V5 housekeeping protocols appears clumsy because its two-level structure is not necessary. The outer envelope address and the inner data link address contain the same information. The two-level structure is needed for ISDN call control because the inner frame structure was previously standardized for ISDN. There could be an advantage to the two-level approach to frame addressing of other protocols if the outer address specified that the message belonged to a specific protocol and the inner address indicated one of a number of different processing entities. However, it is not clear that this would serve any useful purpose, because the routing of protocol messages to processing entities can be performed in any case by a specific implementation.

The multiplexing of the PSTN protocol may cause even greater problems in the future, because it limits the number of PSTN ports and constrains the future use of the PSTN protocol. The number of PSTN ports is limited, because only one time slot is available for PSTN signaling. This time slot may not have sufficient bandwidth to support the 32 thousand PSTN user ports that the V5 interface was otherwise designed to support, since it gives a worst case of 2 bps per user port.

This potential bottleneck on PSTN call control could be removed by allowing the PSTN protocol to operate on several time slots. There is no need for additional frame layer addressing to be introduced to do this, because it is only necessary to specify that all call control information for a given PSTN user port must go through the same time slot. A similar approach could also be adopted if there is a bottleneck due to the activation and deactivation messages on the control protocol or due to the assignment of bearer channels by the BCC protocol, but in these cases changes to the protection protocol would be required, since not all housekeeping protocols would then share the same 64-Kbps time slot.

Using additional frame layer addresses does not in itself remove the potential bottleneck for PSTN signaling, because these addresses are used to differentiate between signaling on the same time slot. Additional frame level addresses could be useful to extend the possible future use of the PSTN protocol. In particular, they could allow PSTN messages to be frame-relayed across the access network between digital telephone sets and the host exchange. The use of the PSTN protocol in this way would allow exactly the same PSTN services to be supported on digital telephone sets as on existing analog telephone sets, which is something that cannot be done by ISDN terminals. It would allow the power required for PSTN service to be reduced, since conventional analog PSTN signaling, particularly ringing, is not low-power.

Modifying the frame layer addressing for PSTN would also allow a rationalization of the V5 approach to addressing. In particular, the need for a the general V5 address space could be eliminated, because all user port addresses could be frame layer addresses. There could also be a rationalization of the control protocol because the message layer address is redundant for the other control functions that are not port-specific.

## 10.7 SUMMARY

The 2.048-Mbps links that make up a V5 interface are formatted into frames of 32 time slots, each of 64 Kbps. Time slot 0 of each link is used to mark the start of each frame, and the each of the remaining 31 time slots is either used for bearer traffic or for communications protocols.

Time slot 16 of the primary link of the interface normally contains the V5 control protocol. For V5.2 interfaces, it also normally contains the other housekeeping protocols, except for the protection protocol, which is transmitted on time slot 16 of both the primary and the secondary 2.048-Mbps links. Following a failure, the protocols that are normally assigned to time slot 16 of the primary link are switched to time slot 16 of the secondary link.

ISDN communications are frame-relayed between the ISDN user ports and the associated V5 communications time slots. Any physical communications time slots, other than those reserved for protection, may be used for ISDN communication or for the V5 PSTN protocol.

There is a general address space used for frame level and message level addressing of V5 communications. In addition, there are addresses specific to the frame and message levels. The different V5-specific protocols and the various ISDN ports are distinguished by different frame level addresses.

The messages of the V5-specific protocols have a generic header plus a message-specific part, both of which are built up from information elements. For convenience, these are described using a convention that allows individual messages to be identified.

The physical layer multiplexing on the V5 interface may create a bottleneck for the PSTN protocol and also for the control and BCC protocols. Removing the bottlenecks for the housekeeping protocols would also affect the protection protocol. There are also peculiarities with the frame layer and message layer addressing used on the V5 interfaces, and with hindsight it is clear that these could have been avoided.

# 11 ISDN Signaling and Multiplexing

"I still don't know."

—Anonymous

The purpose of this chapter is to describe how the signaling at ISDN user ports is multiplexed on the V5 interface. In addition to this basic description, the various other options that were considered are presented and analyzed, together with some of the issues that arose during the creation of the standard.

The physical layer of the V5 interface provides 64-Kbps communications channels that are used by the signaling and control protocols. The only formatting on these channels is the bit numbering, 1 to 8 and this is ignored by the higher protocol layers. The communications channels are treated as if they were simply 64-Kbps bit streams because this allows the well-established layer 2 (the link or frame layer) techniques to be applied.

The principle function of OSI layer 2 is to adapt between the unstructured data stream of the physical layer, which may be corrupted by errors, and the structured layer 3 messages, which are obtained after error correction. Realistically, no form of error correction that is performed can be absolutely reliable, because certain types of error may fool the error correction algorithm. This is not a problem, since unless undetectable errors are introduced deliberately, the error rate would be identified as unacceptably high before there was any significant probability of an undetected error occurring. The probability of an isolated undetected error occurring can be made much less than the probability of other events, such as a multiple hardware component failure or the system being destroyed in a fire.

Formatting and error correction are normally combined, since an error must be localized to a block or frame for it to be corrected. Error correction can be either in-built as forward error correction or it can rely on the detection of errors and a subsequent requested retransmission. In-built error correction is used on the headers of ATM cells, which have an extra octet to enable the location and correction of errors. ISDN call control signaling, on the other hand, assigns a labeling number and error detection octets to each frame so that the corrupted frames can be identified and retransmission requested. Frames for retransmission must be identified with respect to a valid frame, since the label number of corrupted frames is suspect.

Although it is possible to identify particular frames or blocks of frames for retransmission by specifying the valid frames that enclose the corrupted frames, it is simpler to request retransmission from the last valid received frame, since this avoids the need for the receiving end to buffer valid frames and restore the correct frame ordering after the retransmission of previously corrupted frames. The implementation at the transmitting side is also simplified if the receiving side lets the transmitting side know the number of the last valid frame received, since the transmitting side need only store subsequent frames for possible retransmission.

The V5 interface uses an error detection and retransmission approach to error correction taken from CCITT Recommendation Q.921. The V5 approach for ISDN signaling is also designed to take advantage of the low error rate that is expected for access networks by avoiding error correction within the access network.

## 11.1 THE FRAME RELAY CONCEPT

The OSI seven-layer stack does not appear to have been invented by followers of Zen. A cornerstone of Zen is the focusing on reality and the avoidance of misleading conceptual abstractions.

The error correction at layer 2 of the OSI stack is not necessary if there are no errors. If the error rate is very low, then there is no need to perform error correction at every link across a connection, since normally these links are error-free. Error correction need only be applied end to end because this is sufficient and minimizes the processing overhead. This is the basis of the frame relay approach (see Figure 11.1).

## 11.2 THE DEBATE OVER ISDN MULTIPLEXING

It is interesting that the delegates to the initial standards meetings on the V5 interface did not resort to physical violence to resolve the different views about how the ISDN signaling should be multiplexed, and in this respect the debate differed from one on religion, which in other respects it resembled. Three different options were considered, corresponding to OSI layers 1, 2, and 3. The conclusion to use a frame relay approach represented a triumph both of logic and of horse-trading, and afterwards there was a certain amount of drowning of sorrows, but no burning at the stake.

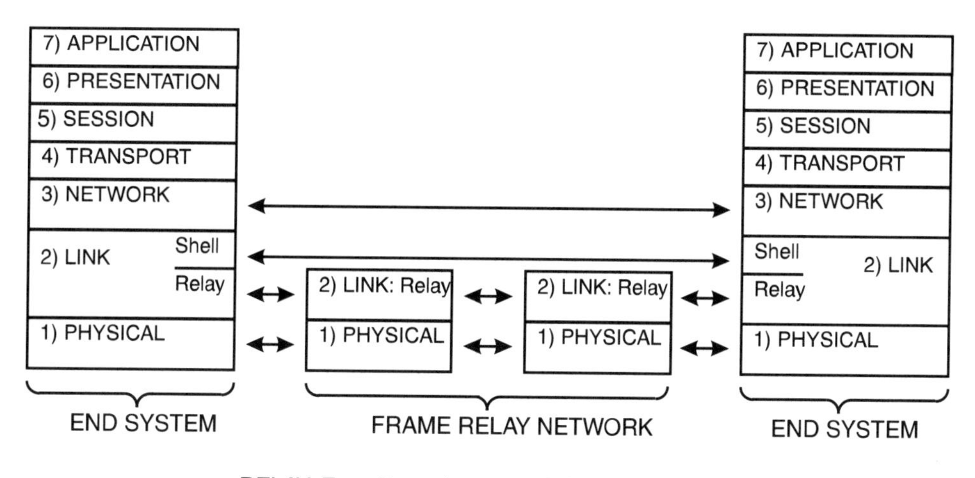


**Figure 11.1** Typical frame relay transmission.

The layer 1 option was based on the simple mapping of basic rate ISDN channels into a 2.048-Mbps link. This would have supported a simple implementation in the access network, because it avoids the processing of the ISDN signaling. One disadvantage of this approach is that unless some form of metasignaling is introduced, it wastes bandwidth on V5 interfaces that concentrate traffic. The bearer channels on a concentrating V5 interface must be allocated to the user ports on demand, and the demand for a bearer channel must be communicated across the V5 interface before the allocation protocol can respond to it. If the demand is identified from the D-channel signaling from the ISDN user port, then there must be a D-channel permanently allocated to each user port. This results in bandwidth being squandered, because bandwidth on the 2.048-Mbps links that could have been allocated on demand to support the B-channels at the user ports must be permanently assigned to D-channels at user ports which may not be active.

This problem with layer 1 multiplexing is not really serious, because bandwidth could also be allocated on demand to support the D-channels at user ports. This allocation could be based on the activation and deactivation signals that normally bracket the flow of traffic (i.e., the activation and deactivation information acts as metasignaling). Ports that are permanently active could have permanently assigned D-channels or be assigned bandwidth quickly from a preallocated pool as soon as signaling is detected.

A more serious problem with the layer 1 option is that it only optimizes the simplicity of the implementation in the access network, not for the interface as a whole. The implementation in the exchange is suboptimal, because it requires D-channel signaling to be handled in parallel, requiring duplication of hardware. This additional hardware can be eliminated if the signaling from a number of ISDN user ports is multiplexed onto a single signaling channel on the V5 interface, and then handled in serial. This form of multiplexing can differentiate between user ports by either layer 2 or layer 3 addresses.

The octets in the frame structures for a layer 3 multiplexing option could be identical to those in a layer 2 option, but the handling of the frame at layer 3 is much more complex. A full layer 3 multiplexing approach requires the layer 3 messages to be terminated and regenerated. Incoming messages should also ideally be checked to ensure that they conform to the specified message format, because the output should not contain invalid messages. In particular, the protocol discriminator should be correct, and the information elements should be of the correct size and value range and correspond to the message type and to the direction of transmission. Layer 3 multiplexing would require the messages to be processed in the access network and allow new messages, for instance, for activation and deactivation, to be added at the V5 interface.

There was a fallacy in the discussion of the layer 3 option in that it was suggested that layer 3 multiplexing would allow a remotely located access network to perform local switching if the interface to the host exchange failed. This is nonsense, because it does not matter what the form of multiplexing is on the interface if the interface itself is not operational. It may be more charitable to assume that this fallacy was introduced deliberately to confuse the issue and help protect a vested interest. If the fallacy had been accepted, it would have forced all implementations to support message processing in the access network. There was a regional requirement to support local switching in the access network if the V5 interface failed, and this does require message processing within the access network. A supplier who had to support this regional requirement would be at less of a disadvantage in comparison with suppliers who did not have to support it, since the adoption of the layer 3 approach would have forced everyone to perform message processing.

It was also suggested that to refuse to terminate layer 3 would bring down the wrath of the OSI seven-layer stack upon the heads of the transgressors. In any event, the layer 3 option was discarded because it introduced unnecessary complexity, and none of the delegates have as yet shown signs of turning into salt.

Layer 2 multiplexing has an advantage over layer 3 multiplexing because it is transparent. This makes it almost completely independent of the nature of the messages it carries. In practical terms, it is immune to upgrades and revisions of the ISDN call control protocol. This is especially important because the cost of upgrading a proprietary layer 3 approach may be higher than the cost of introducing a standard layer 2 approach, especially if the layer 3 approach involves protocol conversion. Similar considerations to those that apply to layer 3 multiplexing also apply to the error correction functions within layer 2 multiplexing. The layer 2 error detection function, however, was retained in the access network because it allows corrupted frames to be discarded and so prevents them from delaying valid frames.

The final agreement to use layer 2 multiplexing of ISDN signaling endorsed the least expensive option to implement, because the interface as a whole is the simplest. It avoids wasting bandwidth for signaling, as in the layer 1 approach. The signaling hardware overall is simpler than for the layer 1 approach, since it can be shared among several ports, rather than having duplicated hardware for each port. It is also simpler than the layer 3 approach, since it does not require both the access network and the exchange to perform message processing.

## 11.3 V5 LAYER 2 MULTIPLEXING FOR ISDN

A valid ISDN signaling frame received from a user port has a port address attached at the start of the frame to indicate the ISDN user port that has sent it and is then passed over the V5 interface to the host exchange. The *frame check sequence* (FCS) at the end of the original frame to allow corrupted frames to be detected is recalculated and the original value is replaced by the recalculated value. The modified ISDN signaling frame from the user port is then sent to the exchange across the V5 interface (see Figure 11.2).

A valid message arriving from the exchange across the V5 interface is processed in the opposite order. The ISDN port address is stripped from the

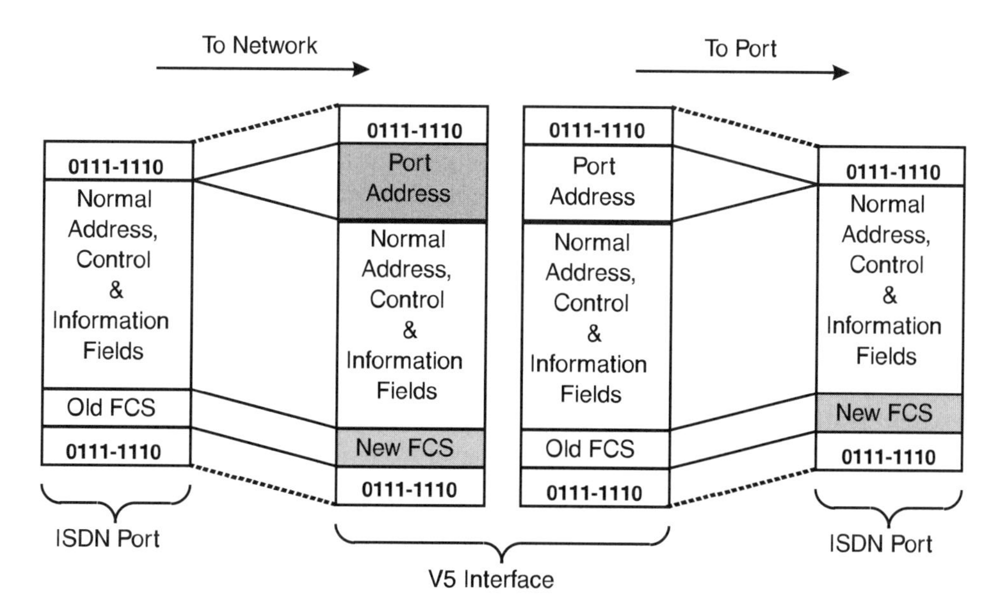


**Figure 11.2** Addition and removal of ISDN port addresses.

message and used to direct the message to the appropriate user port. The FCS for the stripped message is recalculated and replaces the FCS of the original message. The stripped message is then sent across the user port to the ISDN user.

Since this approach is transparent to the layer 3 signaling, the access network does not have to be upgraded if the ISDN signaling is modified, for instance, to include new supplementary services.

## 11.4 ISDN COMMUNICATION PATHS AND CHANNELS

A two-step approach has been adopted for frame-relaying ISDN D-channel information, since the total bandwidth required for ISDN signaling across a V5 interface may be greater than the 64 Kbps available on a single communications channel. The two-step approach also has the potential to simplify the subsequent routing of packets and frames within the telecommunications network. To allow the flexibility necessary to handle differing traffic and routing requirements, user ports are not associated directly with communications channels, but are associated first with communications paths. It is these communications paths that are in turn associated with the communications channels on the V5 interface (see Figure 11.3). This allows subsequent routing within the telecommunication network to be simplified, because different types of D-channel information can be allocated to different communications channels.

Three types of D-channel signaling are identified, corresponding to different layer 2 addresses in the Q.921 recommendation for ISDN signaling. The s-type signaling contains the ISDN call control used to set up and clear down ISDN calls. The p-type and f-type signaling contain user-to-user packet and frame relay communications, respectively, and these will normally be routed to packet and frame relay switches. If there is no need to groom these three types of signaling onto different communications channels, then they can share a single channel, because they are differentiated by their Q.921 layer 2 addresses.

The same type of signaling from a number of different user ports is grouped into a single communications path. There can be a number of s-type, p-type, and d-type communications paths for a V5 interface, and the maximum number of these depends on whether the interface is V5.1 or V5.2. A single type of signaling belonging to a single user port can only be associated with a single communications path of that signaling type, since associating more than one signaling path of the same type with a user port would require some form of coordination of the paths to avoid disrupting the order in which messages are received at the exchange. The approach allows a single user port to be associated with up to three different communications paths, because it may use each of the three different types of ISDN signaling.

A single communications channel can also support up to three different types of ISDN communications paths. There cannot be more than one commu-

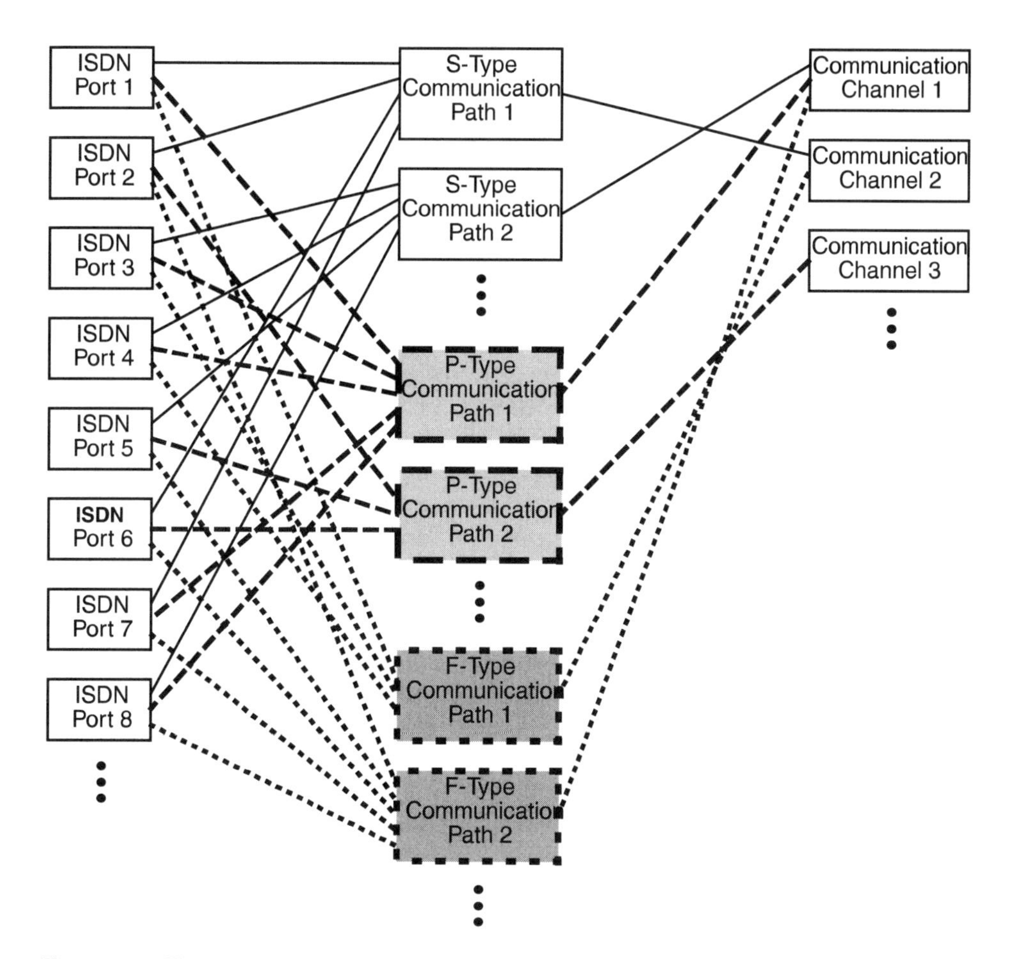


**Figure 11.3** The two-step approach to ISDN multiplexing.

nications path of each type on a single communications channel, because it is impossible to distinguish between different paths of the same type on the same communications channel.

## 11.5 SUMMARY

A frame relay approach is used to multiplex signaling from ISDN user ports onto the V5 interface. This operates at layer 2 of the OSI stack and results in ISDN signaling being transparently multiplexed and demultiplexed by the access network. The detection and retransmission of corrupted frames is performed by the ISDN terminals and by the local exchange, not by the access network.

Terminating layer 3 of the ISDN stack within the access network would have introduced additional complexity, because the access network would then need to interpret ISDN messages. This would create additional complications when ISDN signaling is upgraded, because the access network would also need to be upgraded.

A layer 1 approach to the multiplexing of ISDN signaling would have been conceptually simpler, but would have required additional bandwidth to be dedicated to ISDN signaling on the V5 interface. It would also have required additional hardware to handle each D-channel from each user port. The layer 2 approach allows the same hardware to be time-multiplexed over a number of user ports.

ISDN signaling from user ports is broken down into call control signaling (s-type), user-to-user frame relay data (f-type), and user-to-user packet data (p-type). The same type of signaling from a number of different user ports is multiplexed onto a communications path of that type. Communications paths for different types are in turn multiplexed onto communications time slots. Other protocols may share the same communications time slots as the ISDN communications paths, but different communications paths for the same type of ISDN signaling cannot share the same time slot, because there is no way to distinguish between them.

# 12 The Control Protocol

"What I tell you three times is true."
—Lewis Carroll, *The Hunting of the Snark*

The control protocol is the most important single protocol of the V5 interface, and is the only housekeeping protocol that must always be present. It controls both user ports and other general functions.

The control protocol allows user ports to be blocked and unblocked, and this is used to coordinate the availability of service between the access network and its host exchange. It also allows the access network to inform the host exchange about degradations that might jeopardize service. If the user port is an ISDN port, then the control protocol can be used to monitor and control the activation and deactivation of the port and to prevent the D-channels at user ports from swamping the V5 ISDN communications paths with messages.

In addition to the port-control functions, the control protocol allows checking of the V5 interface identification and configuration, restarting of the PSTN protocol after a failure, and synchronization of the switch-over between the current V5 configuration and a new V5 configuration.

## 12.1 THE FORMAT OF CONTROL MESSAGES

Messages on the V5 interface are identified as belonging to the control protocol by their frame layer address. This takes the value of the control protocol address in the general V5 address space. Specific control protocol messages are then identified by the message type information element in their generic header.

The message type information element identifies only four types of messages for the control protocol (see Figure 12.1). Two of these types, PORT-CONTROL and COMMON-CONTROL, are the originating messages that con-

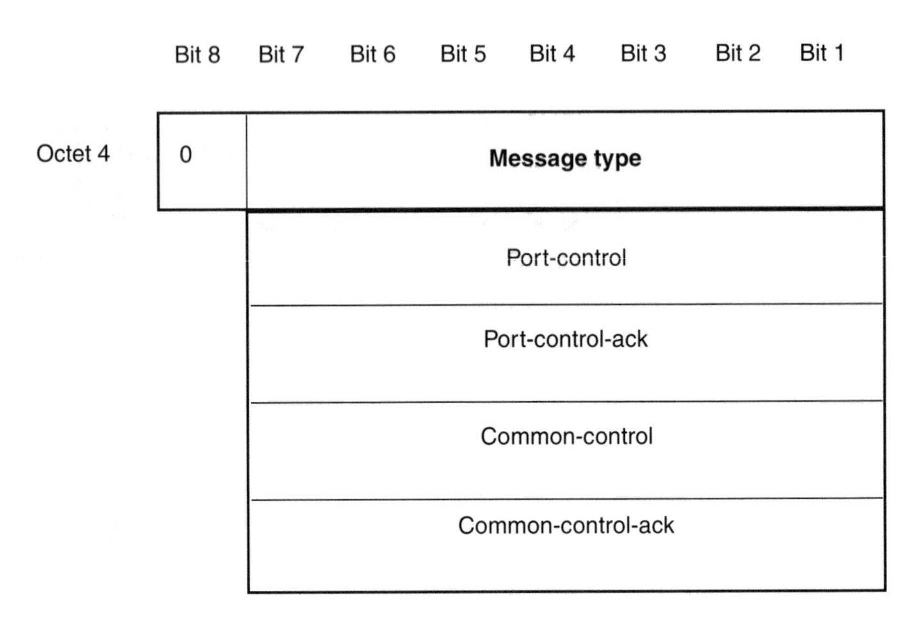


**Figure 12.1** Types of control protocol messages.

trol the functions of the control protocol. The two remaining types of messages, PORT-CONTROL-ACK and COMMON-CONTROL-ACK, are the acknowledgments associated with the two originating messages.

The message address field in the generic header is taken from the general address space. For port-related control messages, the address has the value for the associated PSTN or ISDN port. For common control messages, the address has the value for the control protocol (see Figure 12.2). There is a duplication of information in the header for the control messages, because both the message layer address and the message type information element indicate whether the message is port-related or common control. However, only the message type indicates whether the message is an originating message or an acknowledgment.

The first message-specific information element to follow the generic header is always mandatory, because it identifies the particular function of the originating message or acknowledgment. This information element is the Control-function-element in the PORT-CONTROL and PORT-CONTROL-ACK messages and the Control-function-ID in the COMMON-CONTROL and COMMON-CONTROL-ACK messages.

PORT-CONTROL-ACK messages and PORT-CONTROL messages other than PORT-CONTROL:performance-grading only have mandatory information elements. PORT-CONTROL:performance-grading messages also contain a Performance-grading information element (see Figure 12.3).

| Message | Layer 3 address |
|---|---|
| Port-control | Port address |
| Port-control-ack | Port address |
| Common-control | Control protocol address |
| Common-control-ack | Control protocol address |

**Figure 12.2** Message layer addresses.

| | Message specific information elements | | | | | |
|---|---|---|---|---|---|---|
| **Message** | A | B | C | D | E | F |
| Port-control | M | 1 | | | | |
| Port-control-ack | M | | | | | |
| Comon-control | | | M | 2 | 3 | 4 |
| Comon-control-ack | | | M | | | |

Key:

A = Control-function-element　B = Performance-grading　C = Control-function-D

D = Variant　E = Rejection-cause　F = Interface-ID

**Figure 12.3** Control message-specific information elements. Note: (M) Mandatory; (1) If Control-function-element = performance grading; (2) Not present if Control-function-ID = request-variant-and-interface-ID, or blocking-started, or restart or restart-acknowledge; (3) If Control-function-ID = not-ready-for-reprovisioning or cannot-reprovision; and (4) If Control-function-ID = variant-and-interface-ID.

COMMON-CONTROL-ACK messages also only have a single type-specific information element, the Control-function-ID. Many COMMON-CONTROL messages contain the Variant information element. COMMON-CONTROL:not-ready-for-reprovisioning and COMMON-CONTROL:cannot-reprovision messages also contain a Rejection-cause information element. COMMON-CONTROL:variant-and-interface-ID messages also contain an Interface-ID information element (see Figure 12.3).

The acknowledgment messages are sent in response to the corresponding originating messages and so confirm that the message has been properly received. These acknowledgments are really superfluous, because confirmation of the receipt of the message is already performed at the frame layer. Justification of these messages can be sought on the basis that errors can occur between OSI layers 2 and 3, but this is a general problem which should not be solved on an ad hoc basis.

## 12.2 PORT-CONTROL MESSAGES

PORT-CONTROL messages support the blocking and unblocking of all ports (i.e., PSTN, ISDN, and leased-line ports). The PORT-CONTROL messages also support a number of functions that are specific to ISDN ports, specifically activation and deactivation, fault and performance indications, and flow control for signaling. The various forms of PORT-CONTROL messages are shown in Figure 12.4.

As previously indicated, PORT-CONTROL-ACK messages are used to acknowledge the receipt of the originating PORT-CONTROL message. A corresponding PORT-CONTROL-ACK message is returned for each of the originating PORT-CONTROL messages sent.

### 12.2.1 Port Blocking and Unblocking Messages

Ports on an access network may not be able to support service, either due to a failure or because of maintenance that may have disrupted the service. If service to a port is not possible, either due to a condition within the exchange or within the access network, then a PORT-CONTROL:block message is sent to inform the other side of the V5 interface that this situation has occurred (see Figure 12.5).

The access network is not always aware of whether or not a user port is busy, since ISDN signaling is not interpreted and because some ISDN ports may be active even when there is no signaling or traffic. Because the access network sometimes does not know the status of its user ports, it needs a way to minimize the disruption of service when it wishes to block a port. This is achieved by sending an AN/PORT-CONTROL:block-request message to the exchange. The exchange, which is aware of the status of the port, can then reply with an LE/PORT-CONTROL:block message to indicate that it has blocked the port,

| Value of Control-function-element | Direction (AN:LE) |
|---|---|
| block request | → |
| block | ↔ |
| unblock | ↔ |
| D-channel-block | ← |
| D-channel-unblock | ← |
| activate-access | ← |
| activation-initiated-by-user | → |
| digital-section-activated | → |
| access-activated | → |
| deactivate-access | ← |
| access-deactivated | → |
| performance-grading | → |
| TE-out-of-service | → |
| failure-inside-network | → |

**Figure 12.4** Forms of the PORT-CONTROL message.

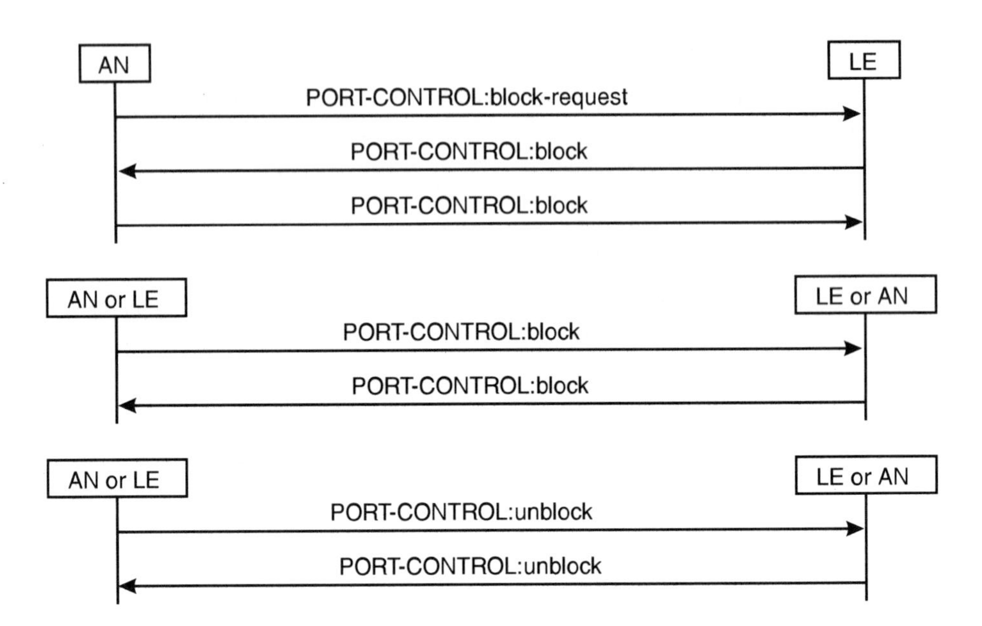


**Figure 12.5** Example message flows for port blocking and unblocking (acknowledgment messages not shown).

either immediately or when the port is no longer busy. The access network may then send its own AN/PORT-CONTROL:block message without the fear of disrupting service.

If the exchange does not respond to the AN/PORT-CONTROL:block-request message within a reasonable time, then the access network may send an AN/PORT-CONTROL:block message and risk the disruption of an ongoing call. The exchange does not need to send a blocking request to the access network, because it can block a port without disrupting an ongoing call, since it is aware of the call state.

To unblock a previously blocked port, both sides must send and receive a PORT-CONTROL:unblock message. Unblocking is aborted if either side sends a PORT-CONTROL:block message or if the access network sends an AN/PORT-CONTROL:block-request message after receipt of a PORT-CONTROL:unblock message.

### 12.2.2 ISDN Flow Control Messages

The concentration of ISDN signaling that occurs on a V5 interface can result in problems, especially if ISDN customers overuse their D-channel signaling, because each basic rate ISDN customer has a 16-Kbps signaling channel at the user port and a large number of customers may share a 64-Kbps channel at the V5 interface. To prevent this, it is necessary for the exchange to be able to ask the access network to block D-channel information before it is statistically multiplexed onto a communications channel, since otherwise the communications channel may become congested with undesired or malicious signaling.

To cut off the D-channel information from an ISDN user port, the exchange sends an LE/PORT-CONTROL:D-channel-block message. To restore the D-channel information from the port, the exchange sends an LE/PORT-CONTROL:D-channel-unblock message.

### 12.2.3 ISDN Activation and Deactivation Messages

The call control protocol for ISDN, which is specified in CCITT Recommendation Q.931 and frame-relayed across the V5 interface, does not contain any information about the activation or deactivation of ISDN ports. This information is relevant to the lower layers of the ISDN transmission and is excluded from layer 3 of the protocol so that the integrity of the layers is preserved.

Support of activation and deactivation signals is required for ISDN basic rate ports, but similar messages are also indirectly associated with primary rate ports if these are supported by a digital section. Messages for activation and deactivation are contained in the V5 control protocol, because they are not included in the frame-relayed ISDN call control messages.

If activation is initiated by the user, then an AN/PORT-CONTROL:activation-initiated-by-user message is sent to the exchange. The user is able to generate this signal by transmitting to the user port. At this point in time, the user will not normally be synchronized to the network, because there is normally no transmission from the network to which the user can synchronize. Typically, the exchange will respond with an LE/PORT-CONTROL:activate-access message, which initiates the transmission of a signal from the network to the user. The user can then extract clock synchronization from this signal and respond to it. This results in an AN/PORT-CONTROL:access-activated message being sent to the exchange. Alternatively, the exchange can initiate activation by sending LE/PORT-CONTROL:activate-access when no AN/PORT-CONTROL:activation-initiated-by-the-user message has been received.

If a digital section is used to support the ISDN port, then the access network sends an AN/PORT-CONTROL:digital-section-activated message when the digital section has become active.

Deactivation may be requested by the exchange through an LE/PORT-CONTROL:deactivate-access message. When the access becomes deactivated, an AN/PORT-CONTROL:access-deactivated message is sent to the exchange.

### 12.2.4 ISDN Port Fault and Performance Messages

The exchange needs to be informed of changes in the performance of a digital section because they can affect the services that the exchange is responsible for.

The access network sends AN/PORT-CONTROL:performance-grading messages to inform the exchange of these changes.

It may be useful for the exchange to know the origin of a failure related to a primary rate port, although just how useful this is has been questioned because it is not clear what the exchange will do with this information or why it should be needed for primary rate accesses but not basic rate accesses. The access network can send AN/PORT-CONTROL:TE-out-of-service to indicate that a failure has been traced to the CPE. Alternatively, it can send AN/PORT-CONTROL:failure-inside-network if the failure is known to be due to an internal access network fault.

## 12.3 COMMON-CONTROL MESSAGES

The COMMON-CONTROL messages make it possible to check that the access network side of one V5 interface has not been connected to the exchange side of a different V5 interface, and that both sides of the V5 interface are consistently configured. They also allow changes in the configuration at either side of the interface to be synchronized and coordinated. The third function of the COMMON-CONTROL messages is to coordinate a general reset of the PSTN protocol, since because of the way in which it has been specified, the PSTN protocol does not do this for itself. The various forms of the COMMON-CONTROL message are shown in Figure 12.6.

Most of the COMMON-CONTROL messages include the provisioning variant label of the sending side. This acts as a constant indication of the interface configuration which exists at the side that sent the message and has been included because both sides of the interface should know the current configuration before attempting to change it. It is not included in the messages used to reset the PSTN protocol, because resetting the PSTN protocol is a separate function from changing the configuration of the V5 interface.

The provisioning variant label is not included in the COMMON-CONTROL:request-variant-and-interface-ID, and this can be understood on the basis of withholding information about the sending side when requesting information about the remote side. Its absence in the COMMON-CONTROL: blocking-started message can also be understood because at this point there is an agreement to change the configuration, but this does not explain its subsequent presence in the COMMON-CONTROL:reprovisioning-started message.

Only COMMON-CONTROL:not-ready-for-reprovisioning and COMMON-CONTROL:cannot-reprovision messages need to give an explanation of what has gone wrong. They use the Rejection-cause information element to do this. Only the COMMON-CONTROL:request-variant-and-interface-ID message identifies the interface ID assigned at the side that sends the message. Other messages do not need to send this information because the interface ID should not

| Value of Control-function-ID | Direction (AN:LE) |
|---|---|
| request-variant-and-interface-ID | ⟷ |
| variant-and-interface-ID | ⟷ |
| verify-reprovisioning | ⟷ |
| not-ready-for-reprovisioning | ⟷ |
| ready-for-reprovisioning | ⟷ |
| switch-over-to-new-variant | ⟷ |
| cannot-reprovision | ⟷ |
| blocking-started | ⟵ |
| reprovisioning-started | ⟶ |
| PSTN-restart | ⟷ |
| PSTN-restart-acknowledge | ⟷ |

**Figure 12.6** Forms of the COMMON-CONTROL message.

need to be checked often, whereas changes to the configuration of an existing interface should be less rare. A mismatch of interface ID between the sides may occur after a physical reconnection, and it is sensible to check the interface ID after the interface has been disconnected. The information is contained in the Interface-ID information element.

COMMON-CONTROL messages are originating messages and are automatically acknowledged with appropriate COMMON-CONTROL-ACK messages. These contain no message-specific information elements other than the Control-function-ID, which, together with their message layer address, indicates the particular COMMON-CONTROL message being acknowledged.

### 12.3.1 V5 Interface Identification

When access networks and exchanges are linked with V5 interfaces, it is sensible to be able to check that the two sides of the interface have been connected correctly. In addition, since a V5 interface can be configured or provisioned in a number of different ways, it is sensible to be able to check that the configurations also match.

Each side of a V5 interface can inform the other side of its interface ID and provisioning variant label by sending a COMMON-CONTROL:variant-and-interface-ID message. It can also request the other side to send such a message to it by sending a COMMON-CONTROL:request-variant-and-interface-ID. Typically, the same interface ID and provisioning variant label will be used on either side of a V5 interface, because they are common to that interface.

### 12.3.2 Reprovisioning of the V5 Interface

Since the V5 interface can support a large number of customers, it is important to minimize the disruption caused by changes to the configuration of the V5 interface. This reprovisioning may be needed because of changes in the traffic across the interface. The V5 interface supports reprovisioning by providing a mechanism to synchronize and coordinate changes to its own configuration. This mechanism can be used at the discretion of the implementation, since the operations systems of the access network and the host exchange may be coordinated in other ways, and the mechanism is more appropriate to major changes than to minor ones, since its additional messages may add unnecessary complexity if the change is small.

When a reprovisioning is performed, it is advisable to ensure that the communications path used by the control protocol is not altered, so that control communication between the two sides of the V5 interface is not lost. This may require the relevant logical communications channel to be switched onto a stable time slot using the protection protocol. Before a reprovisioning is started, it may be sensible to use the V5 interface identification messages (COMMON-CONTROL:request-variant-and-interface-ID and COMMON-CONTROL:variant-and-interface-ID) to confirm that the correct interface is being reconfigured and that the initial configuration has the expected label. Likewise, it may be sensible to use these messages later to confirm that reprovisioning has been successfully completed.

The side of the V5 interface that wishes to initiate a change of configuration should first check that the other side is able to change over to the new configuration (see Figure 12.7). The COMMON-CONTROL:verify-reprovisioning message has been specified for this purpose. When a COMMON-CONTROL:verify-reprovisioning message is received, the response should either be a COMMON-CONTROL:ready-for-reprovisioning message or a COMMON-CONTROL:not-ready-for-reprovisioning message, including the appropriate explanation. If a COMMON-CONTROL:ready-for-reprovisioning message is received, it can be assumed that it is safe to proceed with reprovisioning.

The change to a new configuration is triggered by sending a COMMON-CONTROL:switch-over-to-new-variant message. If the access network triggers the switch-over, then the exchange will typically respond with a LE/COMMON-CONTROL:blocking-started message, block the relevant ports and links, and protect communications channels if required. The exchange can do this with the minimum disruption of ongoing traffic.

Regardless of which side triggers the change, the exchange sends an LE/COMMON-CONTROL:switch-over when it is satisfied that the necessary precautions have been taken. This message ensures that the exchange can control the disruption caused by reprovisioning. This is the first message of the procedure if the exchange triggers switch-over, because the exchange can take the preliminary precautions without seeking permission from the access network.

The side that receives a COMMON-CONTROL:switch-over message may be unable to comply with the request. In this case, it responds with a COMMON-CONTROL:cannot-reprovision message explaining what the problem is. The access network must receive the LE/COMMON-CONTROL:switch-over message indicating that the exchange is satisfied before proceeding with reprovisioning. To proceed, it sends an AN/COMMON-CONTROL:reprovisioning-

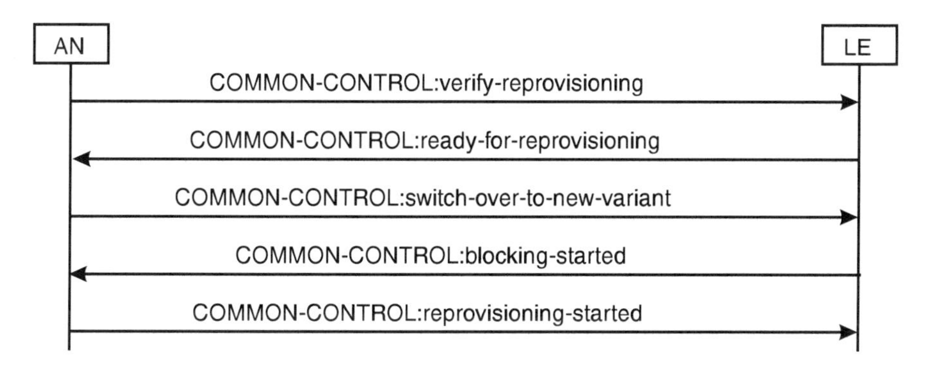


**Figure 12.7** Example message flow for reprovisioning initiated by the AN (acknowledgment messages not shown).

started message and is then free to make the change. The exchange is free to make the change after it receives this message.

Once the change is completed, ports and links should be unblocked and communications channels that have been temporarily switched to stable time slots may be restored. There are no additional messages to indicate that reprovisioning has been completed. Additional messages are unnecessary, because the existing messages can be used. A COMMON-CONTROL:variant-and-interface-ID message can be used to indicate that the new configuration has been installed and a COMMON-CONTROL:request-variant-and-interface-ID message can be used to check the version installed at the opposite side.

It may be necessary to confirm that reprovisioning is not yet complete at the opposite side. This can be done by sending a COMMON-CONTROL:verify-reprovisioning message. If reprovisioning is not yet complete, then a COMMON-CONTROL:not-ready-for-reprovisioning/Cause:reprovisioning-in-progress message can be returned. If the process is completed, then a COMMON-CONTROL: not-ready-for-reprovisioning/Cause:variant-known-not-ready message should be returned to indicate that reprovisioning is complete and no further reprovisioning is required.

The reprovisioning procedures specified for the V5 interface are flawed, because they are not sufficiently clear as to how, when, and why the new provisioning label should be used, both during the initiation of the change and to confirm its completion. The procedures also do not cover how a reprovisioning can be aborted if a problem is discovered and the original reprovisioning restored. Perhaps the most enlightened aspect of the reprovisioning procedure is that its use is optional.

### 12.3.3 Restart of the PSTN Protocol

It is possible under unusual conditions that the PSTN protocol may need to be reset to its initial state. This is a greater problem for the PSTN protocol than for the other V5 protocols, because the messages of the PSTN protocol are mapped onto events at user ports that take no account of the need for initialization of the protocol. A separate reset function in the V5 control protocol has been defined, because there are no generic PSTN protocol messages, only messages that refer to specific PSTN ports. V5 protocols other than the PSTN protocol support whatever alignment and resetting functions they require by themselves.

If either side of the V5 interface wishes to reset the PSTN protocol, it issues a COMMON-CONTROL:restart message. This is acknowledged by the receiving side, which must send back a COMMON-CONTROL:restart-acknowledge message. Both COMMON-CONTROL:restart and COMMON-CONTROL: restart-acknowledge are originating messages and so are acknowledged by COMMON-CONTROL-ACK:restart and COMMON-CONTROL-ACK:restart-acknowledge messages, respectively. The combination of error correction and acknowledgments at the frame layer, plus two levels of handshaking at the message layer, can be defended on the basis that resetting the PSTN protocol can affect several thousand customers, so this ensures that both sides are fully committed to performing the restart.

## 12.4 CONCLUDING REMARKS

There are a number of aspects of the V5 control protocol that appear peculiar or questionable. In many cases, these have been included for reasons that are not technical, but this is often the case with international standards. It is important to identify and understand these aspects to avoid propagating technically poor solutions and to counteract the common misconception that international standards are sacrosanct.

The use of the PORT-CONTROL-ACK and COMMON-CONTROL-ACK messages is perhaps the most questionable aspect of the specification of the V5 control protocol. The confirmation that messages have been delivered is already built into the frame layer procedures through the control fields and the frame check sequences. Frames are numbered and checked to ensure that they are error-free. If errors are found, then retransmission from the last correctly received frame is requested.

It is not appropriate to add additional acknowledgments at the message layer for PORT-CONTROL messages, because this level of security is not required. The probability of an undetected failure at the frame layer is less than the probability of disruption by other events, and if this situation did occur, then only one port would be affected. It is also not appropriate to use the acknowledgments for common control functions, because message level checking already exists. The messages used for coordinated reprovisioning already ensure that there is extensive checking and handshaking before reprovisioning is started. There is also an acknowledgment message separately defined for the PSTN restart, and the PSTN restart procedure would not normally be initiated unless a serious fault with the PSTN protocol had already been discovered and corrective action taken.

The procedure to block and unblock ports is not consistent with CCITT Recommendation X.731 on the states used for management. This causes problems when attempting to use the approaches of the widely accepted CCITT Recommendations on *telecommunications management network* (TMN). Matching these with the V5 approach results in greatly increased complexity on the management interfaces.

The use of the performance monitoring messages is suspect because they are limited to ISDN ports with remote digital sections. There is no reason to assume that a remote digital section is either more or less reliable than the rest of the transmission system between the user and the exchange. All user ports should be treated equally regardless of remote digital sections, because the same considerations apply equally.

There is an inconsistency with respect to the fault location messages for ISDN PRA ports. These messages allow the access network to inform the exchange if a fault is within the access network or if it is associated with the customer equipment. The requirements for this are questionable; in particular, it is not clear why this is needed for PRA ports supported by the V5.2 interface, but not for basic rate ports supported by the V5.1 interface.

There is also an inconsistent approach towards the control of other protocols. The reset function for the PSTN protocol is included in the control protocol, although it could have been included in the PSTN protocol in the same way that the reset function for the protection protocol is built into that protocol directly. On the other hand the PROTOCOL-ERROR message of the protection protocol could also be used for other protocols and should therefore either have been built into them all individually or been included as a common control function in the control protocol.

These peculiarities of the V5 control protocol are not fundamental flaws. Rather, they are a manifestation of the nontechnical factors that influence an international standard. The fact that it has been possible to reach agreement at all on the V5 standard between people with different cultural backgrounds and with different requirements is a tribute to human cooperation and stands in contrast to much of human history.

## 12.5 SUMMARY

A control protocol is required to support the activation and deactivation signals required for ISDN ports. It is also necessary to control the flow of messages from ISDN ports, so that users are prevented from swamping the V5 communications channels, and to indicate problems that affect service to the exchange. It must be possible to block and unblock individual ports of every kind if faults occur on these or if maintenance is required.

The control protocol must be able to check the identity of the V5 interface to ensure that interfaces are correctly interconnected. The control protocol also allows changes to the configuration of both sides of the interface to be coordinated and the PSTN protocol to be reset.

A system of originating messages and associated acknowledgments is used to implement the control protocol. There are only four types of control messages. These are the originating port and common control messages and the related acknowledgment messages for port and common control. The specific functions are identified by values of an information element within each message.

A number of aspects of the control protocol appear peculiar. The use of acknowledgment messages is questionable, because the confirmation that messages have been properly delivered is already performed by the frame layer. The approach used for blocking and unblocking of ports is not consistent with CCITT Recommendation X.731 on state management, and this causes problems for the management of the interface. The use of port-related fault and performance messages is also questionable, because these messages are equally applicable to ports without remote digital sections. It is not clear why the ISDN PRA fault location messages are required, and yet the corresponding basic rate messages are not. There is also an inconsistency between the inclusion of the PSTN protocol reset and the exclusion of the protection protocol error message. These peculiarities do not prevent the operation of the control protocol, but there is no reason to propagate them in future specifications.

# 13 The PSTN Protocol

"If I called the wrong number, why did you answer the phone?"
—James Thurber

The PSTN protocol is described in more detail than the other V5 protocols because it is more complex and more flexible than the other protocols. The PSTN protocol also has the potential to be used in new applications where the ISDN call control protocol is inappropriate, perhaps due to the difficulty of simultaneously mapping a number of variants of PSTN onto ISDN. Examples of potential applications are digital PSTN interfaces and signaling for video services where otherwise ad hoc protocols could be adopted due to expedience and then become entrenched.

The requirements for a generic call control protocol are discussed first to clarify the aim of the protocol. Then the relationship of these requirements with the information elements and messages of the V5 PSTN protocol is described. The PSTN protocol is in effect a call control toolkit. It can be used in a variety of different ways to meet the requirements of different operators. The way in which this toolkit is used for a given application is specified in the mapping between the PSTN signals and the PSTN protocol messages, and this mapping will normally be different for different countries, because they have different histories for PSTN signaling. An example of a simple mapping is included as an illustration.

## 13.1 THE PSTN PROBLEM

Contrary to many people's expectations, the work on the V5 interface specification has proven that it is not possible to map PSTN call control onto ISDN call control. It may be possible in some cases for a single country or operator to map their specific PSTN call control onto ISDN, but not always. Even worse, fundamental inconsistencies were discovered when a number of different national approaches to PSTN call control had to be simultaneously mapped onto ISDN call control without compromising any national PSTN service. Those who think that this is a trivial ISDN application problem are advised to attempt the solution in detail for the PSTN requirements of several different countries.

The PSTN protocol also differs from ISDN because the PSTN layer 3 messages originate and terminate in the access network. In the ISDN case, the messages can be frame-relayed to the user port without being interpreted by the access network. For PSTN, processing is performed at the message layer, since PSTN ports cannot handle layer 3 signaling messages, and so the access network must translate between the PSTN protocol and the signals at the user port.

However, the solution used for PSTN makes use of the same principle of transparency that was used for ISDN signaling. The access network does as little processing of the PSTN signals as possible. A stimulus approach is used and the finite state machine for the call state is kept very simple. The detailed understanding of PSTN call control is the responsibility of the host exchange, not the access network.

Stimuli generated by the PSTN terminal (e.g., off-hook, on-hook) are mapped directly into PSTN protocol messages, and a message layer address is added to identify the PSTN port. These messages are then transmitted to the exchange over the V5 interface (see Figure 13.1). Messages received from the exchange are examined for the message layer address used to identify the user port for which the message is intended. The message is then translated into the appropriate stimulus (ringing, line reversal), and this is applied to the PSTN port.

This approach makes no assumptions about the sequence of messages and the response of the exchange. Although different operators have been unable to

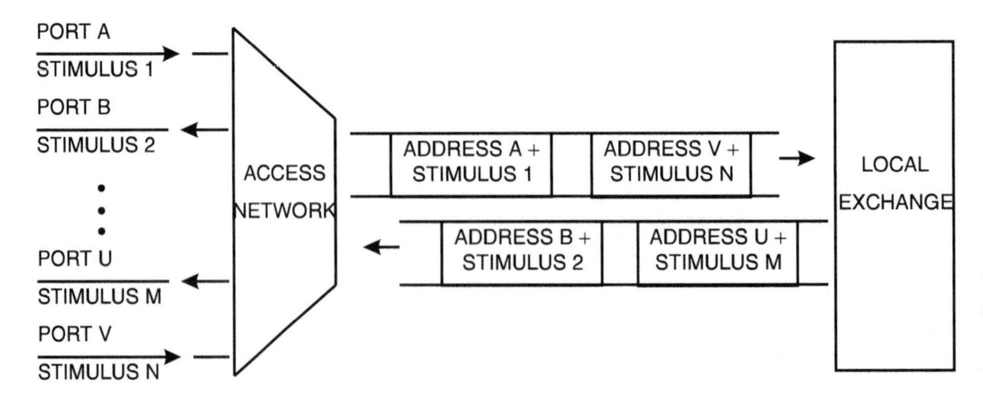


**Figure 13.1** The stimulus approach to PSTN signaling.

agree on the sequences and responses for PSTN, they have been able to agree on the stimuli that need to be conveyed across the V5 interface.

A certain amount of additional complexity is required for PSTN call control because it has not been possible to prevent the presence of an access network from having some effect. In particular, using an OSI stack across the V5 interface and the queuing and processing of messages themselves introduce additional delays. In order to meet various time constraints on PSTN call control, it is necessary to compensate for these delays.

## 13.2 BASIC CALL CONTROL REQUIREMENTS

The main function of a call control protocol is setting up and clearing down connections. However, this is not the only requirement for the protocol. Information relating to the connection, such as charging information, must also be sent both during the connection and after the connection has terminated. In addition, information about the condition of the line also needs to be sent, because this can also affect the way in which connections are handled, and these conditions can change at any time.

Information about the condition of the line is normally available at the exchange if no access network is present. Likewise, information about the circumstances of the PSTN call are normally available at the line circuit. Messages are required when an access network is present between the customer and the exchange, because information that would be available if the user port were directly connected to the exchange must be conveyed.

The second-order effects of having a call control protocol also need to be taken into account. These are due to the delays introduced by the processing of the messages and by the need to handle anomalous or unexpected messages.

### 13.2.1 Call Phases and Call Cycle Points

There are four underlying states involved in making a call. Two of these are the transitional states of setting-up or clearing-down a call. Between the two transitional states, there are the two stable states: the busy state after call setup and before call clear-down and the quiescent phase after call clear-down and before any subsequent call setup. This is illustrated in Figure 13.2.

The two best defined events of the call cycle are the start of the setting-up state and the end of the clearing-down state, because there can be ambiguity about the start and end of the busy state. For example, a call to a remote answering machine could be considered to become busy once the connection is set up. However, if the functionality of the answering machine is implemented by an intelligent telecommunication network, then no end-to-end con-

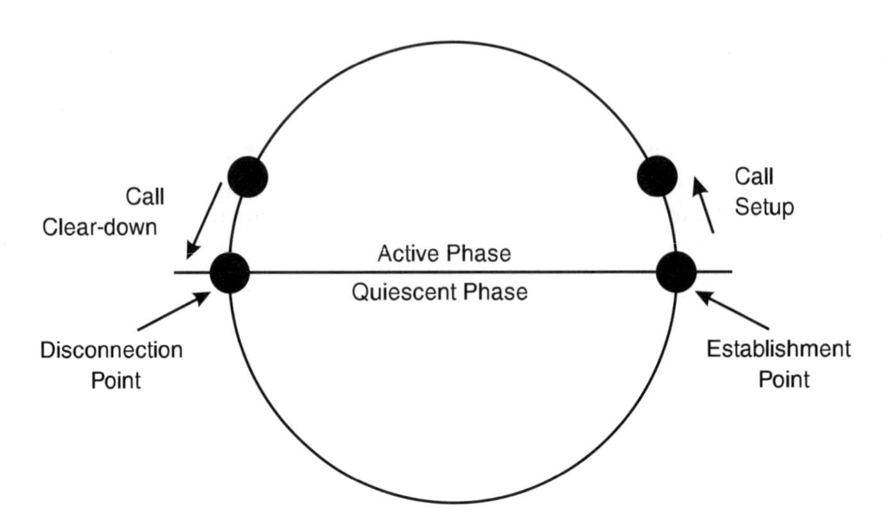


**Figure 13.2** Call phases and call cycle points.

nection need be established, and it could be claimed that there is no busy state, since the call is never completed.

The two points at either end of the quiescent state will be referred to as the *establishment* and *disconnection points*, because this clarifies the relationship with the corresponding messages of the V5 PSTN protocol. To avoid ambiguity, the active phase can then be defined as the period from the establishment point to the disconnection point. During the active phase, various call control signals may be exchanged. The quiescent phase is just the period between the disconnection point and the next establishment point.

### 13.2.2 Line Information

There is a requirement in PSTN to inform the exchange about the creation or removal of a specific line condition (e.g., a 30-k Ω loop resistance), which can be used to activate or deactivate call diversion. If no access network is present, then the exchange can sense this line condition directly. A call control message is required across the V5 interface, because the exchange no longer has the ability to directly sense the line when an access network is present.

This line condition is normally controlled by a switch on the CPE. It should be possible to inform the exchange of changes in the position of the switch whenever its position is altered. This is not a problem during the active phase of a call because the information can be transmitted as one of the call control signals. Difficulty arises if the call diversion switch is altered during the quiescent call phase, because no call control signals occur in this phase. Under these conditions, a special type of call relating to line information must b briefly set up and then cleared down to convey the state of the line.

There are other line conditions the exchange would be able to sense directly if no access network were present. These can be handled in the same way, either by a call control signal during the active phase of a normal call, or by a special call if the condition occurs during the quiescent phase.

### 13.2.3 Protocol Delays

The use of a call control protocol for PSTN introduces additional delays for which the protocol itself must be able to compensate. In particular, the PSTN protocol must ensure that response times already defined for PSTN implementations are maintained when the PSTN service is delivered using an access network.

The protocol may need to allow the exchange to enable specific responses to be automatically returned by the access network, because the protocol is otherwise unable to meet the necessary response times. This allows a spontaneous reflex response to be made by the access network. Alternatively, the protocol may need to allow the exchange to activate predefined procedures in the access network, so that fast responses can be made with the exchange still retaining a level of direct involvement.

### 13.2.4 Protocol Anomalies

The creation of a call control protocol allows the possibility of protocol anomalies to occur. In some cases, they may be the result of unforeseen situations occurring in the telecommunications network. In other cases, they may be the result of software bugs, different interpretations of the protocol specification, or other unforeseen interworking difficulties.

The protocol needs to have appropriate mechanisms to allow these anomalies to be flagged. In particular, the exchange must be kept aware of anomalies in the access network that affect service, because the exchange is responsible for this.

## 13.3 PSTN STIMULI AND INFORMATION ELEMENTS

The messages of the PSTN call control protocol consist of the general header for the protocol plus the message-specific information elements that represent the PSTN stimuli or additional information (see Figure 13.3). The general header identifies the message as a V5 protocol message, identifies the port to which it refers, and identifies the point of the call with which it is associated through the message type. Many of the stimuli can be used at different points in a call and

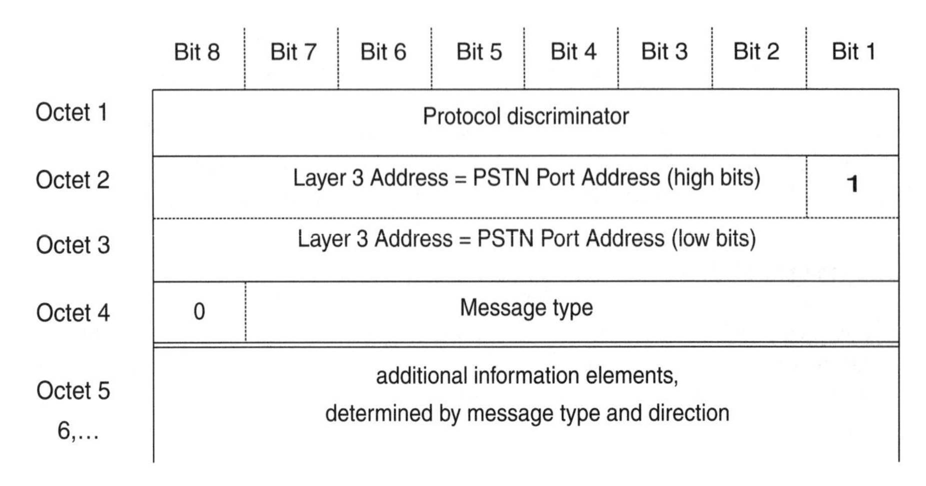


**Figure 13.3** PSTN message format.

have different effects. For instance, the effect of dialing a digit is typically quite different after a connection has been established. Before the various PSTN messages are described, the requirements for the various specific information elements that are grouped together to form the messages are examined.

### 13.3.1 Fundamental Stimuli and Information Elements

For most of the time, PSTN handsets can be described simply as on-hook or off-hook. Likewise, for most of the duration of a call, the polarity of the telephone line is fixed. Both of these situations can be described as steady conditions, signaled either by the access network or by the exchange. These and other similar conditions are indicated in the Steady-signal information element of the V5 protocol (see Figure 13.4).

It is not efficient to use the separate steady-state messages that indicate on-hook and off-hook to signal make-break dialing pulses, since this would require a large number of messages to be generated to convey a complete telephone number, and the messages would probably need to be time-stamped to prevent variations in message processing delays from interfering with the recognition of the dialed digits. An alternative approach is to perform the recognition of make-break dialed digits in the access network and to indicate the identified digit in a signal. The V5 Digit-signal information element is used for this. It may be sent in either direction because it is used for both normal dialing sent to the exchange and *direct dialing-in* (DDI) signals sent by the exchange to PABXs.

| | Information Element |
|---|---|
| Fundamental | Steady-signal |
| | Digit-signal |
| | Cadenced-ringing |
| | Pulsed-signal |
| | Pulse-notification |
| Specialized | Line-information |
| | Recognition-time |
| Delay Compensation | Enable-autonomous-acknowledge |
| | Disable-autonomous-acknowledge |
| | Autonomous-signaling-sequence |
| | Sequence-response |
| Maintenance Housekeeping | Sequence-number |
| | State |
| | Cause |
| | Response-unavailable |

**Figure 13.4** Information elements for PSTN.

One of the most common signals to be sent by the exchange is the ringing signal. This is indicated by the Cadenced-ringing information element. This information element allows a number of different ringing types to be identified, so that it is possible to represent personalized ringing.

In certain circumstances it is necessary for the exchange to indicate that metering pulses should be sent to the customer's terminal. There is also a requirement for the access network to indicate that a recall pulse has been detected by the access network. Pulsed information of this kind is conveyed in the Pulsed-signal information element. In addition, a Pulse-notification information element may be used to inform the exchange about the successful generation of pulses, because it may be necessary for the exchange to have confirmation that the correct metering pulses have been sent to the customer.

### 13.3.2 Specialized Information

The four fundamental information types (steady, dialed digits, ringing, and pulses) cover most of the basic requirements of the PSTN call control protocol. These types of information are likely to be required in any PSTN implementation. In addition to these there are specialized types of information that may not be needed in every case.

The need to convey information to the exchange about the line condition has already been mentioned. This information is contained in the Line-information information element, which can be used to activate or deactivate call diversion. This information element is not universally required because there are alternative approaches to the control of call diversion.

It may also be necessary to change the time for which a signal must be present before it is deemed valid. This is not universal requirement, but it is definitely needed, for instance, when the recognition time needs to be changed to reduce the chance of misinterpreting the line condition. This requirement can be met by using the Recognition-time information element.

### 13.3.3 Delay Compensation Elements

The PSTN protocol requires signaling information to be assembled into messages transmitted in a channel that is shared by all PSTN ports. When the messages arrive, they are checked and then deciphered. This introduces a variable processing delay between an event being detected in the access network and the access network responding to the event following the interchange of messages with the exchange.

In order to avoid this delay, the exchange can request the access network to respond to a particular signal with a specific identified response. The signal and its associated response are specified in an Enable-autonomous-acknowledge information element. The exchange can also cancel this automatic response by the access network with a message containing a Disable-autonomous-acknowledge information element.

The automatic responses the access network can provide may be predefined in the access network if the implementation is simple. However, it may also be necessary for the access network to react according to a sequence of signals that would normally require several PSTN messages. Sequences that are predefined, in contrast to responses that are pre-defined, are not identified explicitly in a message but are indicated instead by a label in a message. The sequence to which the label refers is determined by the implementation. Predefined sequences of signaling may be activated by the Autonomous-signaling-sequence information element. There is no corresponding deactivation information element, but if deactivation is required, then it could be achieved by activating a sequence that is predefined as an overriding deactivation instruction.

If the access network needs to send a response to the exchange relating to a predefined sequence, then this response is given using the Sequence-response information element.

### 13.3.4 Maintenance Information Elements

The PSTN protocol needs to perform certain maintenance tasks on itself, and there are a number of information elements associated with these tasks.

Certain of the messages associated with call control need to be numbered to help with the detection of errors. The Sequence-number information element is used to label these messages. It may be used in either direction of transmission.

An error-free call control message may be received that does not make sense in the context of the other messages that have been received. If this happens, then clarification is needed about the state of the call. This clarification is given in the State information element, and the reason for sending this is given in the associated Cause information element.

It is also possible for the protocol to be used to make a request that cannot be carried out. For example, the protocol may be used to request the generation of metering pulses at a port in the access network that does not support metering pulses. The requesting message may be correctly numbered, and the message may make sense within the context of the other messages exchanged. The Resource-unavailable information element can be used to indicate that the request is not supported.

## 13.4 CALL CONTROL MESSAGES

The information elements of the call control protocol are used within the various call control messages sent between the access network and its host exchange. There are nine messages in the call control protocol, and these are associated with the different parts of the call cycle (see Figure 13.5).

At the establishment point of the call cycle, there is an exchange of ESTABLISH and ESTABLISH-ACK messages. Likewise, at the disconnection point of the cycle, there is an exchange of DISCONNECT and DISCONNECT-COMPLETE messages. During the active phase, between these points, SIGNAL and SIGNAL-ACK messages are exchanged. The exchange may also adjust the behavior of the access network by sending a PROTOCOL-PARAMETER message during the active phase.

During any part of the call cycle, the exchange can send a STATUS-ENQUIRY message if it receives a message that is out of context. The access network will send a STATUS message at any part of the call cycle if it receives a message that is out of context or if it receives a STATUS-ENQUIRY message from the exchange.

### 13.4.1 Call Establishment Messages

The quiescent phase of the call cycle is ended by either the access network or the exchange sending an ESTABLISH message (see Figure 13.6). The access

| | Bit 8 | Bit 7 | Bit 6 | Bit 5 | Bit 4 | Bit 3 | Bit 2 | Bit 1 | |
|---|---|---|---|---|---|---|---|---|---|
| Octet 4 | 0 | **Message type** | | | | | | | **Direction (AN:LE)** |
| | | Establishment Point | | ESTABLISH | | | | | <-> |
| | | | | ESTABLISH-ACK | | | | | <-> |
| | | Active Phase | | SIGNAL | | | | | <-> |
| | | | | PROTOCOL-PARAMETER | | | | | <-- |
| | | | | SIGNAL-ACK | | | | | <-> |
| | | Disconnection Point | | DISCONNECT | | | | | <-> |
| | | | | DISCONNECT-COMPLETE | | | | | <-> |
| | | Any Time | | STATUS-ENQUIRY | | | | | <-- |
| | | | | STATUS | | | | | --> |

**Figure 13.5** Call control messages.

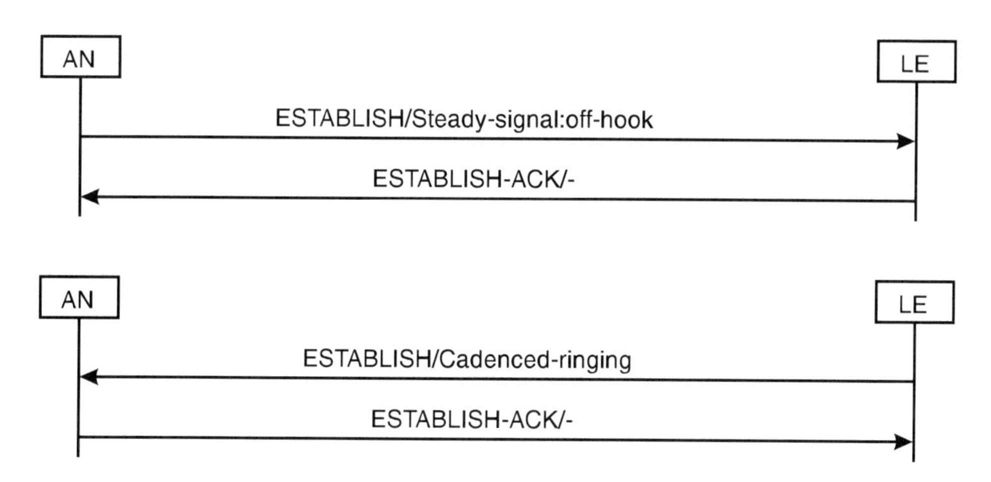


**Figure 13.6** Example message flows for ESTABLISH messages.

network may send an AN/ESTABLISH/- message if no additional call control information is to be sent at this time. Alternatively, the access network can send additional fundamental call control information about a new steady state by using an AN/ESTABLISH/Steady-signal. The new steady state could be off-hook, which might otherwise be sent in a subsequent AN/SIGNAL message if it is not included in the initial message. The specific information elements for call establishment are shown in Figure 13.7.

If the exchange wishes to end the quiescent phase and has no need to send any additional call control information, then it sends an LE/ESTABLISH/- message. The exchange can also send additional fundamental call control information concerning ringing and steady or pulsed conditions. An LE/ESTABLISH/ Cadenced-ringing message can be used to send normal ringing, while an LE/ ESTABLISH/Pulsed-signal can be used to send an initial ringing pulse. An LE/ESTABLISH/Steady-state signal can be used to define battery conditions to be applied to the line. If ringing is not requested in the LE/ESTABLISH message, it may need to be requested in a subsequent LE/SIGNAL message.

Alternatively, the exchange may activate a predefined signaling sequence in the access network by sending an LE/ESTABLISH/Autonomous-signaling-sequence message. The inclusion of this delay compensation element allows the access network to signal to the customer and to respond to the customer without needing to wait for responses from the exchange.

The access network can respond to an LE/ESTABLISH message with an AN/ESTABLISH-ACK/Steady-state message to indicate, for example, that the customer has responded to the incoming call by going off-hook. Alternatively,

| Message | Message Specific Information Elements | | | | |
|---|---|---|---|---|---|
| | B | D | G | H | J |
| AN/ESTABLISH | 1 | | | | 1 |
| LE/ESTABLISH | | 1 | 1 | 1 | 1 |
| AN/ESTABLISH-ACK | | | | 1 | 1 |
| LE/ESTABLISH-ACK | | 1 | | 1 | 1 |

KEY:
B = Line-information D = Autonomous-signaling-sequence
G = Cadenced-ringing H = Pulsed-signal J = Steady-signal

**Figure 13.7** Message-specific information elements for call establishment. Note: (1) Optional, but only one may be used.

the access network could respond with an AN/ESTABLISH-ACK/Pulsed-signal message to request special functions to be applied to the incoming call. If no additional call control information needs to be sent, then the access network responds with a simple AN/ESTABLISH-ACK/- message.

The exchange can respond to an AN/ESTABLISH message with an LE/ESTABLISH-ACK/Steady-signal message, for instance, to request specific battery conditions to be applied to the customer's line. Alternatively, the exchange can respond with an LE/ESTABLISH-ACK/Pulsed-signal, for instance, to request metering pulses to be sent to the customer, or the exchange can respond with an LE/ESTABLISH-ACK/Autonomous-signaling-sequence to activate a predefined response sequence in the access network. If no additional information needs to be sent, then the exchange can send a simple LE/ESTABLISH-ACK/- message.

The call establishment messages described above are exchanged when connections are to be made. If the purpose of call establishment is to inform the exchange of a change of line conditions when there is no normal call in progress, then the access network sends an AN/ESTABLISH/Line-information message. This terminates the quiescent phase of the call cycle, but the normal active phase has no duration. In particular, the exchange does not acknowledge this message with an LE/ESTABLISH-ACK message, but with an LE/DISCONNECT message. This is described in more detail in the section on call disconnection messages.

### 13.4.2 Active Phase Messages

There are three types of messages that are specific to the active phase of the call cycle. SIGNAL messages may be sent by either the exchange or the access network to convey a variety of information, and PROTOCOL-PARAMETER messages may be sent by the exchange to alter the response of the access network to signals from the customer. These messages are acknowledged by SIGNAL-ACK messages, which are the third type of message.

To ensure that the sequence of messages is guaranteed, all three types of messages contain a Sequence-number information element. In the SIGNAL and PROTOCOL-PARAMETER messages, this is used to label messages with the order in which they are sent. In the SIGNAL-ACK message, this is used to indicate the number of the next message expected and to confirm that all previous messages have been received correctly. There is an advantage in this case in having an acknowledgment at the message layer, despite the fact that similar acknowledgments are used at the frame layer, because it keeps the message layer software informed about the progress of delivered messages. The SIGNAL-ACK message has no other function and so need not be described further. SIGNAL and PROTOCOL-PARAMETER messages contain a single additional information element that is determined by the type of information conveyed. The specific information elements for these messages are shown in Figure 13.8.

AN/SIGNAL messages, sent by the access network, can include any of the fundamental types of information elements, except Cadenced-ringing, since the access network cannot cause ringing in the exchange. AN/SIGNAL/Steady-state messages can be used to indicate new on-hook or off-hook conditions. AN/SIGNAL/Digital-signal messages can be used to convey dialed-digit information to the exchange (see Figure 13.9). AN/SIGNAL/Pulsed-signal messages can be used to indicate a recall pulse to the exchange, and AN/SIGNAL/Pulse-notification messages can be used to inform the exchange of the successful sending of metering pulses to the customer.

LE/SIGNAL messages, sent by the exchange, can include any of the fundamental information types except Pulse-notification, since pulse-notification is only used by the access network to confirm the delivery of pulses to the customer so that the exchange can be certain that the correct number of metering pulses has been delivered. The application of normal ringing can be requested by the exchange through an LE/SIGNAL/Cadenced-ringing message. An LE/SIGNAL/Pulsed-signal message can precede this to generate an initial ringing pulse, or it can be used to send metering pulses to the customer. The exchange can also dial inwards to a PABX using LE/SIGNAL/Digit-signal messages.

The exchange cannot use LE/SIGNAL messages to send maintenance information, since only the exchange needs to be informed of potentially service-affecting problems at the other side of the interface, but it can use LE/SIGNAL messages for delay compensation. Through a LE/SIGNAL/Autonomous-signal-

| | Message Specific Information Elements | | | | | | | | | | | |
|---|---|---|---|---|---|---|---|---|---|---|---|---|
| **Message** | A | D | E | F | G | H | J | K | L | M | N | P |
| AN/SIGNAL | 2 | | 2 | M | | 2 | 2 | 2 | | | | 2 |
| LE/SIGNAL | | 2 | | M | 2 | 2 | 2 | 2 | | | | |
| AN/SIGNAL-ACK | | | | M | | | | | | | | |
| LE/SIGNAL-ACK | | | | M | | | | | | | | |
| LE/PROTOCOL-PARAMETER | | | | M | | | | | 2 | 2 | 2 | |

Key:
A = Pulse-notification
D = Autonomous-signaling-sequence
E = Sequence-response
F = Sequence-number
G = Cadenced-ringing
H = Pulsed-signal
J = Steady-signal
K = Digit-signal
L = Recognition-time
M = Enable-autonomous-response
N = Disable-autonomous-response
P = Resource-unavailable

**Figure 13.8** Message-specific information elements for the active call phase. Note: (M) Mandatory; and (2) One and only one of these elements must be included in each message.

ing sequence, the exchange can activate a predefined signaling sequence. The access network can indicate the outcome of a predefined signaling sequence through an AN/SIGNAL/Sequence-response message. Of course, the access network may send other messages following the activation of a predefined signaling sequence, but the signaling sequence can make it unnecessary to wait for a response from the exchange.

Unlike the exchange, the access network can use a SIGNAL message to send maintenance information. It can use an AN/SIGNAL/Resource-unavailable message to identify a request from the exchange that cannot be carried out because the access network is not equipped to do so.

Although the exchange cannot use a SIGNAL message to send maintenance information, it can send maintenance information using a PROTOCOL-PARAMETER/Recognition-time message. This message allows the exchange to alter the time taken to recognize signals from the customer, depending on the state of the call. The exchange can also use PROTOCOL-PARAMETER messages for delay compensation. It can enable and disable the automatic generation of specific responses to specific customer signals using LE/PROTOCOL-PARAMETER/Enable-autonomous-acknowledge and LE/PROTOCOL-PARAMETER/Disable-autonomous-acknowledge messages. The access network cannot send any PROTOCOL-PARAMETER messages, since it is the exchange, and not the access

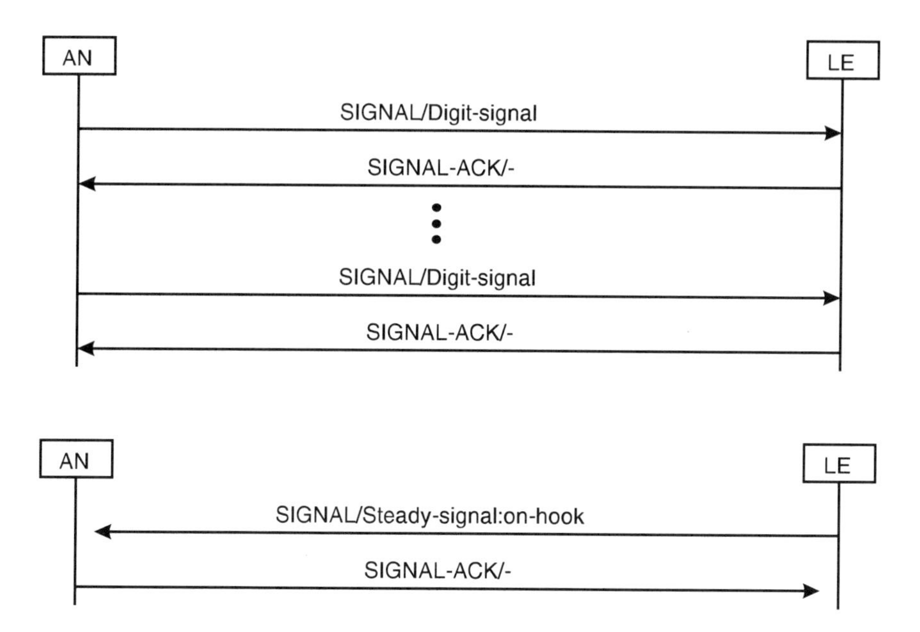


**Figure 13.9** Example message flows for the active call phase.

network, that is responsible for changing the parameters and arming autonomous responses.

### 13.4.3 Call Disconnection Messages

Disconnection ends the active phase of a call and marks the start of the quiescent phase. Disconnection is normally initiated by the exchange, because the exchange, not the customer's equipment, controls disconnection for PSTN, although disconnection by the exchange may be triggered by the customer's telephone going back on-hook (see Figure 13.10). The exception to this rule is an aborted call, since an aborted call has no active phase.

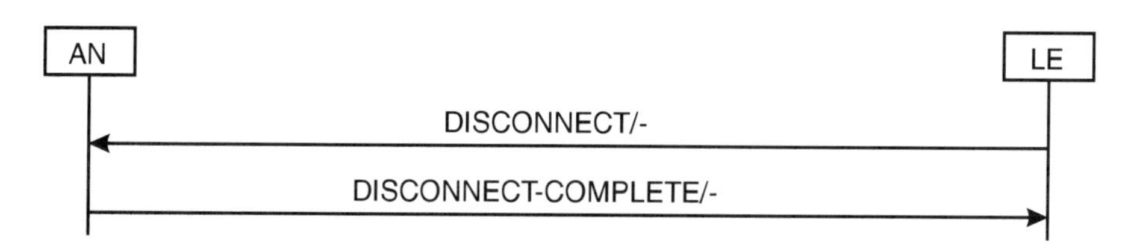


**Figure 13.10** Example message flow for disconnection messages.

The only type of additional information that the exchange can include in a DISCONNECT message is fundamental information about the steady state. The LE/DISCONNECT/Steady-signal message can request a return to a specific steady-state condition for the quiescent phase. The exchange can alternatively use an LE/DISCONNECT/- message, because the steady state for the quiescent phase may be predefined, making identification of it unnecessary. The specific information elements for disconnection are shown in Figure 13.11.

The access network returns an AN/DISCONNECT-COMPLETE message to the exchange to indicate compliance with the DISCONNECT message from the exchange. This message contains no additional information, and so cannot tell the exchange whether the customer is on-hook or off-hook. This is not a problem if the disconnection is in response to an AN/SIGNAL/Steady-state:on-hook message indicating that the line is now idle or parked. If the call is not cleared locally in this way, then some other approach may be required so that the exchange can decide against requesting ringing later when the customer has remained off-hook.

After the access network receives a DISCONNECT message, the call cycle enters its quiescent phase. For the exchange to be informed that the customer has gone on-hook, an AN/ESTABLISH message of some sort can be used. This is required because no messages are normally sent during the quiescent phase and ESTABLISH messages are used to terminate the quiescent phase. The access network could signal a subsequent on-hook using an AN/SIGNAL/Line-information message, but this is not recommended because it is really a misuse of the information element. An AN/SIGNAL/Steady-state: on-hook message is a more natural choice, and the exchange could respond with a DISCONNECT message, possibly preceded by an ESTABLISH ACK message. Although the access network could indicate the return to a quiescent state using a STATUS message, this is not appropriate because no messages that are out of context have been received. If the customer remains off-hook for a sufficient length of time, then this could be deemed to be a

| Message | Message Specific Information Elements |
|---|---|
| DISCONNECT | Steady-signal (optional) |
| DISCONNECT-COMPLETE | Steady-signal (optional) |

**Figure 13.11** Message-specific information elements for disconnection.

further call request and an AN/ESTABLISH/Steady-signal:off-hook message could be sent.

If a call is aborted, then the access network sends an AN/DISCONNECT message to the exchange on receipt of the LE/ESTABLISH-ACK message. An AN/DISCONNECT/Steady-signal:on-hook message could be used to indicate that the customer has gone back on-hook. Alternatively, an AN/DISCONNECT/-message could be sent if no additional information is thought necessary. The exchange completes the abort sequence by sending an LE/DISCONNECT-COMPLETE/Steady-signal message or an LE/DISCONNECT-COMPLETE/-message, depending on whether or not it is necessary to request a specific steady state for the quiescent phase.

### 13.4.4 Status Messages

Status messages are required for the checking of the alignment of call cycles between the access network and the exchange and can be useful in fault situations to realign call states at either side of a V5 interface. This checking can be triggered by the arrival of a message that is out of context, because this is likely to indicate that the call cycles are no longer aligned.

The description of the call cycle given so far needs to be refined, because it is too simple for the clarification of misalignments. Normally, the call cycles in the access network and in the exchange are aligned, so the distinction between them has been ignored till now. In addition, the V5 protocol contains three different types of call cycle (normal, aborted, and line-condition), which have not been properly differentiated till now.

Aborted call cycles are initiated by the customer, but then aborted before a connection is established. Line-condition call cycles also terminate without an end-to-end connection being established, and are used to inform the exchange of a change of line conditions. Both of these abnormal call cycles are initiated by a message from the access network. Normal call cycles can be initiated by either the access network or the exchange, and may result in an end-to-end connection being established.

The quiescent state in the access network corresponds to the "null[1]" state (see Figure 13.12). This means that in the V5 protocol it is called the *access network null state* and referred to as AN1. All call cycles start from the "null[1]" state and finish by returning to it.

Normal call cycles involve the "null[1]," "initiated[2]," and "active[5]" states. Aborted call cycles never reach the "active[5]" state, but instead involve the "aborting[3]" and "disconnecting[7]" states. Line-condition call cycles only involve the "null[1]" and the "info[4]" states.

The exchange may initiate a normal call cycle by sending an LE/ESTABLISH message. The access network responds by sending an AN/ES-

| Inital State | Final State | | | | | |
|---|---|---|---|---|---|---|
| | **NULL 1** | **INITIATED BY AN 2** | **ABORT REQUESTED 3** | **LINE INFO 4** | **PATH ACTIVE 5** | **DISCONNECT REQUESTED 7** |
| 1 | | | | | | |
| 2 | DISCONNECT COMPLETE | | | | ESTABLISH ACK | |
| 3 | DISCONNECT COMPLETE | | | | | |
| 4 | DISCONNECT COMPLETE | | | | | |
| 5 | DISCONNECT COMPLETE | | | | | |
| 7 | DISCONNECT | | | | | |

Typical Unacknowledged PSTN State Transitions for the Access Network

**Figure 13.12** Access network call states for PSTN. Entries indicate associated triggering messages received from host exchange. Other states entered due to local events at the access network.

TABLISH-ACK message and immediately enters the "active[5]" state (see Figure 13.13).

The access network may initiate a normal call by sending an AN/ESTABLISH message and going into the "initiated[2]" state. On receipt of an LE/ESTABLISH-ACK message, the access network goes into the "active[5]" state.

A normal call cycle is terminated when the access network receives an LE/DISCONNECT message. The access network responds by returning to the "null[1]" state and replying with an AN/DISCONNECT-COMPLETE message.

The V5 protocol defines an aborted call cycle to handle the situation where a normal call cycle need not be completed. An aborted call cycle starts as a normal cycle initiated by the access network, but is then aborted before the exchange responds with an LE/ESTABLISH-ACK message. If an LE/ESTABLISH-ACK message is received before the abortion of the cycle is complete, then the "path_active[5]" state is entered and the normal call cycle must be completed.

The replacement of the handset causes the access network to enter the "aborting[3]" state. The subsequent receipt of LE/ESTABLISH-ACK then results in the access network sending an AN/DISCONNECT message and entering the "disconnecting[7]" state. The AN/DISCONNECT message would typically indicate the on-hook condition due to the replacement of the handset. The exchange responds by returning an LE/DISCONNECT-COMPLETE message, which causes the access network to return to the "null[1]" state.

The abort cycle eliminates the need for an AN/SIGNAL/Steady-signal:on-hook message and associated LE/SIGNAL-ACK message. It also avoids the need for the exchange to control an aborted call. However, an implementation could avoid the use of an abort cycle by ignoring the effect of the replacement of the handset until the LE/ESTABLISH-ACK message is received.

A line-condition call cycle starts with the sending of an AN/ESTABLISH/Line-information message. On sending this, the access network enters the

| Stimulus | Response | | | | |
|---|---|---|---|---|---|
| | ESTABLISH | ESTABLISH ACK | SIGNAL | DISCONNECT | DISCONNECT COMPLETE |
| Line Seize | 1 ⟶ 2 | | | | |
| Line Info. | 1 ⟶ 4 | | | | |
| Line Signal | | | 5 ⟶ 5 | | |
| ESTABLISH | | 1,2,3 ⟶ 5 | | | |
| ESTABLISH ACK | | | | 3 ⟶ 7 | |
| DISCONNECT | | | | | 1,2,3,4,5 ⟶ 1 |

Typical PSTN Stimulus and Response for the Access Network

**Figure 13.13** Typical state transitions and responses. Entries indicate associated AN state transitions.

| Message | Message Specific Information Elements |
|---|---|
| STATUS-ENQUIRY | (none) |
| DISCONNECT-COMPLETE | State and Cause (both mandatory) |

**Figure 13.14** Message-specific information elements for status messages.

"info[4]" state. On receipt of the responding LE/DISCONNECT message, the access network returns an AN/DISCONNECT-COMPLETE message and returns to the "null[1]" state.

In addition to the states associated with the call cycles, there are also two service states ("blocked[6]" and "out-of-service[0]"). An AN\STATUS message will also indicate these conditions.

The purpose of status messages is to ensure that the exchange is informed of the status of the call cycle in the access network. The exchange needs to be aware of this because it is responsible for the service, whereas the access network is not. In the V5 protocol, the status of the call cycle in the exchange can only be inferred, while that of the access network can be directly observed.

STATUS messages are only sent by the access network, and STATUS-ENQUIRY messages are only sent by the exchange (see Figure 13.14). These messages only contain mandatory information elements. There are no message-specific information elements in STATUS-ENQUIRY messages. The only message-specific information elements in the STATUS messages are Cause and State.

The reason for sending an AN/STATUS message is given in its Cause information element. This reason may be that an LE/STATUS-ENQUIRY message has been received. The state of the access network, from "out-of-service[0]" to "disconnecting[7]," is indicated in the State information element.

## 13.5 EXAMPLE OF THE USE OF THE PSTN PROTOCOL

There are a large number of ways in which the PSTN protocol can be used, and a particular way will normally be chosen for each specific national implementation of the V5 interface. In the simple example that follows, information elements that are fixed by the message's context are ignored, although they must still be present. SIGNAL-ACK messages are also ignored because they are completely fixed by their context, but likewise must also be present.

Exceptional situations are ignored because the purpose here is to give an example aimed at introducing the concepts, not to give a comprehensive specification designed to meet the specific implementation required for some country.

### 13.5.1 Establish Messages

- AN/ESTABLISH/Steady-signal:off-hook—Used to initiate an outgoing call and triggered by the customer going off-hook.
- LE/ESTABLISH/Cadenced-ringing:0—Used to initiate an incoming call and apply ringing when there is no collision between incoming and outgoing calls.
- LE/ESTABLISH/Steady-signal:normal-polarity—Used to initiate an incoming call when there is a collision and priority is given to the incoming call, since this situation is treated as if ringing had been applied and been answered by the customer going off-hook.

### 13.5.2 Establish Acknowledgment Messages

- AN/ESTABLISH-ACK/-
- LE/ESTABLISH-ACK/-

These two messages are generated automatically on receipt of ESTABLISH messages.

### 13.5.3 Signal Messages

- AN/SIGNAL/Digit-signal:value+no-acknowledgment—Generated when the access network detects digits dialed by the customer.
- AN/SIGNAL/Steady-signal:off-hook—Generated when the customer goes off-hook for a reason other than to initiate an out-going call (e.g., in response to incoming ringing).
- LE/SIGNAL/Steady-signal:normal-polarity—Generated when the exchange wishes to stop ringing in response to an off-hook signal from the access network.
- LE/SIGNAL/Steady-signal:stop-ringing—Generated when the exchange wishes to stop ringing for reasons other than in response to an off-hook signal from the access network.
- AN/SIGNAL/Steady-signal:on-hook—Generated in response to the customer going on-hook unless the on-hook is part of a dialed digit, or unless an LE/DISCONNECT/- message has been received.

### 13.5.4 Disconnect Messages

- LE/DISCONNECT/- —Generated when the exchange wishes to terminate the call, and results in the port being returned to its null state.
- AN/DISCONNECT/- —Generated if the customer goes back on-hook before an LE/ESTABLISH-ACK/- is received in response to an AN/ESTABLISH/Steady-signal:off-hook message.

### 13.5.5 Disconnect Complete Messages

- AN/DISCONNECT-COMPLETE/-
- LE/DISCONNECT-COMPLETE/-

These two messages are generated automatically on receipt of DISCONNECT messages.

## 13.6 SUMMARY

The PSTN protocol is effectively a call control toolkit that can be used in a variety of ways to meet the requirements of different operators and different countries. It has been defined because it is impossible to specify a single mapping of PSTN call control onto ISDN call control that simultaneously meets all the requirements.

The processing of PSTN signals and the state machine for PSTN has been kept as simple as possible, with most events at the PSTN user port being mapped simply onto PSTN protocol messages. The main function of the protocol is the setting up and clearing down of connections. The protocol must also be able to convey information about the condition of the line that can be used for call diversion. Additional messages are also required to compensate for the delays introduced by the protocol itself, and to signal anomalous conditions.

The protocol has protocol-specific information elements that can be classified into four categories: fundamental, specialized, delay compensation, and maintenance.

The protocol uses certain information elements to convey four types of fundamental signaling conditions. These allow steady conditions, pulsed conditions, dialed digits, and ringing to be signaled. There are also two types of specialized information elements used to indicate special changes of the line condition and to change the recognition time for signals.

The protocol also uses other information elements to compensate for the message processing delays introduced by the protocol itself. These control automatic responses by the access network and activate predefined sequences of signaling in the access network. The smooth operation of the protocol itself is ensured by additional maintenance information elements.

The specific messages that make use of the four categories of information elements are used at call establishment, call disconnection, and during the active phase of a call. Status messages can be sent whenever one of the other messages is received out of context.

A simple example of how the different messages can be mapped onto PSTN signals is given.

# 14 The Bearer Channel Connection Protocol

"If a gi'e her any more she'll blow, captain."
—Scotty, *Star Trek*

The BCC protocol has the potential to revolutionize the structure of the telecommunications network, because it allows the exchanges to be extremely small or few in number. A major exchange could become as small as a PABX, since the racks of line cards normally present could be replaced with a few V5.2 interfaces. The BCC protocol allows the physical size of the V5 interface to be significantly reduced, which in turn reduces its cost significantly.

The messages in the BCC protocol control the association between the 64-Kbps channels at the user ports and the 64-Kbps bearer time slots on the V5.2 interface. These messages allow bearer time slots to be allocated to user ports when they are required, and to be deallocated when service is no longer needed. By allocating the time slots to the user ports according to the demand, the protocol allows the interface to concentrate traffic (i.e., for there to be fewer bearer time slots on the V5 interface than there are channels at the user ports). Service is not noticeably affected by this, because only a fraction of the user port channels are active at any one time.

In addition to the allocation and deallocation of bearer time slots, the BCC protocol provides related capabilities. In particular, it allows the host exchange to check that the allocation is correct, and it allows the access network to inform the exchange of faults that may affect the allocation.

## 14.1 ALLOCATION, CONCENTRATION, AND RELIABILITY

A V5.1 interface is not capable of controlling the relationship between the bearer time slots on the interface and the corresponding channels at the user ports, because the association is either fixed by the hardware of the access network or controlled by configuration messages over a separate management interface. This approach is adequate for a V5.1 interface because it does not support concentration and because it cannot survive the failure of one of its links, since it only has one link.

Concentration allows the interface to share its bearer time slots between a larger number of channels at the user ports. User ports do not need to be permanently connected to the time slots on the interface because they are inactive for most of the time. When a user port no longer needs a time slot on the interface, the time slot can be given to another user port that has a use for it. This allows the traffic from a large number of lightly used user ports to be concentrated onto a smaller number of heavily used time slots at the interface. Although this increases the complexity of the interface, the interface becomes much smaller and the cost of the interface per supported user port is much less.

The V5.1 interface cannot support concentration because it does not allow user ports to be given time slots when they need them. To support concentration, a mechanism must be defined to allow the dynamical allocation of time slots according to the demand. This mechanism must be controlled by the exchange, because the access network is not always aware of which ports require time slots, since it does not interpret the call control signaling.

Concentration requires the dynamic allocation of time slots, but the dynamic allocation of time slots does not imply that there must be concentration. For concentration to be present, the access network must also be designed so that there are more bearer channels at the user ports than there are bearer time slots at the interface. When concentration is used, there are occasions when a call may be blocked because there is no bearer time slot available to carry the traffic.

Paradoxically, an interface can become more reliable if it is designed so that it does not guarantee that bearer connections can always be made. On rare occasions, a concentrating interface may not be able to complete a bearer connection, because all the bearer time slots are in use. This makes for a more reliable interface under normal conditions, because bearer connections for all user ports can still be made, even if one of the links fails. The concentration itself does not increase reliability directly, but the dynamic allocation needed to allow concentration to work allows bearer connections to avoid links that have gone down.

## 14.2 REQUIREMENTS FOR THE BCC PROTOCOL

The BCC protocol must be able to handle the allocation of bearer channels at the start of a call and the deallocation of the channels at the end of a call. It should also allow the allocation of channels to be checked so that inconsistencies can be identified, and it should allow faults that affect connections to be reported.

The allocation of bearer channels at the user ports to bearer channels on the V5.2 interface is less complex than in a future ATM network, because all bearer channels are identical. If bearer channels had additional attributes, for instance, different data rates or quality-of-service characteristics, then these would have to be specified in addition to the mapping between the ports and the interface.

The access network cannot initiate the allocation or deallocation of bearer channels because it does not process the control signaling. The access network must be able to reject an allocation that is requested by its host exchange, since the exchange might erroneously make inappropriate requests, such as the allocation of a bearer channel at a user port to a signaling channel on the V5.2 interface. It is also necessary for the access network to be able to confirm that it has successfully completed the allocation or deallocation, because the layer 2 confirmation that the request message has been received is not the same as a confirmation that the requested operation has been performed.

There is no need to identify bearer channels at a PSTN user port, because there is only one channel. For ISDN ports it is necessary to identify the particular B-channel or channels for a specific allocation or deallocation. The allocation or deallocation of channels at an ISDN primary rate port should be possible en bloc, because otherwise up to 30 messages might be needed in rapid succession. It is important for allocations to be able to override an existing allocation so that new allocation with high priority can still be made if the interface becomes congested, although this could also be done by deallocating a low-priority connection first. It is also sensible to be able to abort an incomplete allocation, since this avoids the need to wait until an allocation is complete before it can be deallocated.

### 14.2.1 Allocation and Deallocation of V5.2 Bearer Channels

Once the host exchange has identified that a connection should be made to a bearer channel at a user port, it must indicate this to the access network so that the access network can be made aware of the need. It is better for the host exchange to also indicate the desired bearer channel on the V5.2 interface, because this reduces the amount of subsequent negotiation. If this V5.2 bearer channel is also acceptable to the access network, then the access network can simply perform the requested allocation and then indicate that it has done so. If the access network is not prepared to perform the requested allocation, then it can reject it, preferably giving the reason for the rejection.

In order to perform a deallocation, it is sufficient to indicate either the bearer channel at the user port or the bearer channel at the V5.2 interface, since the connection to be removed can be identified by either of its ends. There is some merit in identifying both ends of the connection, since this ensures that any confusion is detected and connections are less likely to be removed by mistake. Deallocations must be initiated by the host exchange, since the access network does not handle call control. If the exchange sends an invalid deallocation request, such as one where the two ends do not correspond to an existing connection, then the access network should reject it. Otherwise the access network should comply with the request and send an indication to the exchange when it has completed the necessary tasks.

### 14.2.2 Auditing of Connections

The BCC protocol requires complementary mappings of bearer channels in the access network and in the exchange. It is necessary to have some means of checking these mappings so that any inconsistencies can be discovered. For this, it is not sufficient to use a deallocation message that specifies both ends of the connection to reveal inconsistencies, because this could cause the access network to destroy the connection, and this may not be appropriate. If inconsistencies are revealed by a deallocation message, or even by an allocation message, then this may act as a trigger that causes the bearer channel connections to be audited.

If the host exchange specifies both ends of the connection in an allocation request, then the access network should provide explicit information about the connection in response to an audit request, since the exchange already knows what the connection should be. When the exchange sends an audit request to the access network, it should only specify one end of the connection, because this forces the access network to examine its own internal mapping of connections in order to respond to the request. It should be possible for the exchange to specify either the end at the user port or the end at the V5.2 interface, since this makes tracing connections easier.

It should not be necessary to perform auditing of all connections in the access network, because several hundred connections may exist at a V5.2 interface and most of these are likely to be valid. The minimum requirement is to be able to audit connections individually, since the entire set of connections can then be audited one at a time. It would also be useful to audit all connections at a particular user port or V5.2 link, but this introduces additional complexity in the messages without introducing additional functionality, since each of the bearer channels could be checked individually. Although there is a slight overhead in the number of messages if auditing is only performed on individual connections, this is not significant, since auditing of large numbers of connections should not be required very often in normal operation.

### 14.2.3 Handling of Faults and Errors

Connections within an access network that have been established by the BCC protocol may be broken when a fault occurs or due to other events within the access network. If this happens, then the host exchange must be informed, since it is likely that service, which is the responsibility of the exchange, has been affected. A fault message from the access network need not indicate both ends of the affected connection in order to identify it, since a connection can be identified by just one of its ends; but as for a deallocation request from the exchange, there is an advantage if both ends are specified, because this allows inconsistencies to be detected. Unlike a deallocation message, a fault message may not be able to give both ends of the connection because a fault may cause this information to be lost. It is not essential for fault indications to be explicitly acknowledged by the exchange, since the receipt of the message is confirmed by the lower protocol layers.

Protocol errors that relate to the BCC protocol may be detected by either side of the V5.2 interface, but it is more important for the exchange to be informed of errors detected by the access network, because these may indicate that it is necessary for the exchange to carry out an audit of the connections within the access network. The delivery of a protocol error message is confirmed by the lower protocol layers and so need not be explicitly acknowledged. Ideally, any protocol errors flagged within the BCC protocol should refer only to the BCC protocol, and errors that are not specific to the BCC protocol should be flagged using some other mechanism. Unfortunately, there is sometimes an ad hoc approach to the handling of protocol errors in protocol specifications, since this can easily be overlooked and then added as an afterthought without the benefit of careful thought or guiding principles.

### 14.2.4 Common Message References

It is more effective if messages that are related can indicate this through a common reference that does not refer to a particular connection or its end points, because this does not depend on the specification of a valid connection. The use of a common reference also allows acknowledgments to be kept simple.

Related messages can be marked as belonging to the same process through a label specified by the initial message of the process. Subsequent related messages can use the same label to indicate that they are associated with the initial message. Messages belonging to the same process may refer to an invalid connection or to a connection that was originally identified by one of its end points. After the last message of a process, the reference label can then be freed for later use by a new initial message.

The use of process labels simplifies the performance of operations in parallel, since the messages for different processes can be interspersed without confusion. This can also be achieved if some other means is used to indicate how the interspersed messages are associated, but the use of an existing label that was defined for some other reason can have undesired implications and nuances that may lead to confusion.

## 14.3 BEARER CHANNEL CONNECTION MESSAGES

The message layer addresses of the messages of the BCC protocol are used to refer to the BCC process to which the message relates (see Figure 14.1). This makes it unnecessary for the messages to explicitly identify both the V5.2 time slot and the user port channel.

Processes are deemed by the exchange to have terminated on the receipt of a completion or rejection message or if there is no response from the access network after the initial message of the process has been sent for a second time. The lower layers of the protocol stack ensure that the receipt of messages is always actively acknowledged, so the access network should not get confused if the exchange later reuses the process reference for a new process.

Processes are normally initiated by the exchange, since the exchange handles the call control signaling, but they may also be initiated by the access network. All BCC messages contain a BCC-reference-number information element to indicate the process to which they refer. In addition to the reference number itself, this information element also indicates whether the process has been initiated by the access network or by the exchange using the S-bit (see Figure 14.1). The same value of reference number could be used by a process

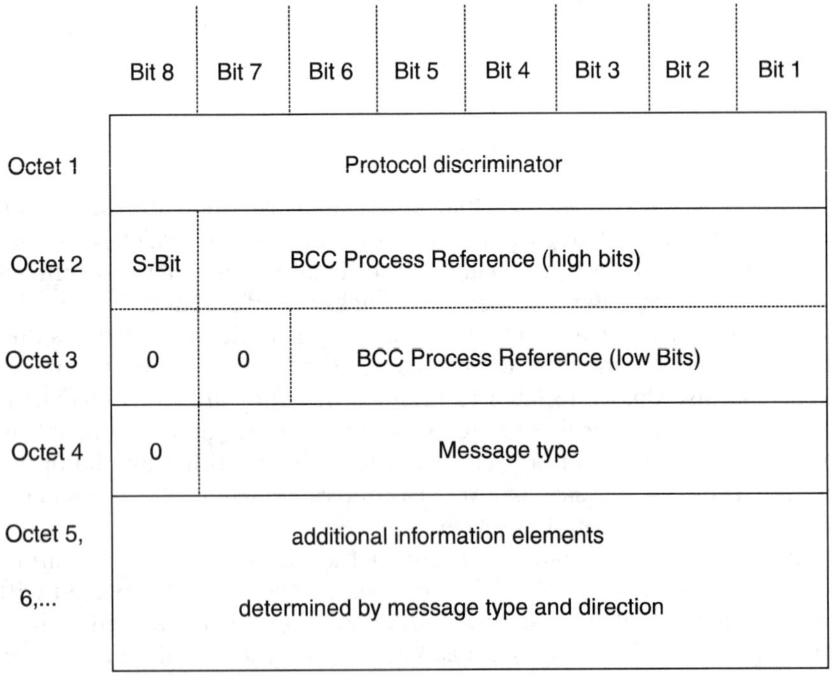

| | Bit 8 | Bit 7 | Bit 6 | Bit 5 | Bit 4 | Bit 3 | Bit 2 | Bit 1 |
|---|---|---|---|---|---|---|---|---|
| Octet 1 | Protocol discriminator | | | | | | | |
| Octet 2 | S-Bit | BCC Process Reference (high bits) | | | | | | |
| Octet 3 | 0 | 0 | BCC Process Reference (low Bits) | | | | | |
| Octet 4 | 0 | Message type | | | | | | |
| Octet 5, 6,... | additional information elements determined by message type and direction | | | | | | | |

**Figure 14.1** BCC protocol message format.

initiated by the exchange and by one initiated by the access network, and although this situation is unlikely, the confusion it would create must be avoided.

The local exchange acts as the master for the BCC protocol and the access network acts as the slave, because the access network does not know when bearer channels are needed, since it does not perform call control. The BCC protocol treats every allocation or deallocation of a V5.2 bearer channel as a separate process, identified by a separate process reference number. Each process terminates with the successful completion of the allocation or deallocation or with the abortion of the process. Different processes can occur in parallel so that a problem with an allocation or deallocation does not delay other allocations and deallocations. The types of messages of the BCC protocol are shown in Figure 14.2.

### 14.3.1 Allocation Messages

The local exchange requests the allocation of a V5.2 time slot by sending an ALLOCATION message to the access network (see Figures 14.3 and 14.4). This contains a new process number (in the mandatory BCC-reference-number information element) used in later messages that refer to this allocation. It also contains the layer 3 address for PSTN ports or the envelope address for ISDN ports (in the mandatory User-port-identification information element), which labels the port on a V5 interface. For PSTN ports, the channel at the user port does not need to be identified, because there is only one. For single-ISDN-bearer-channel allocations, the channels at the user port are identified by the ISDN-port-channel-identification information element, while for multiple-bearer-channel allocations they are identified by the Multi-slot-map information element.

For single-bearer-channel allocations, either for PSTN or for ISDN ports, the V5.2 time slot is identified by the V5-time-slot-identification information element. This indicates both the V5.2 link and the time slot on that link. The information element also specifies whether or not any existing connection should be overridden so that high-priority connections can immediately usurp existing connections.

For multiple-bearer allocations for ISDN, the Multi-slot-map information element, which identifies the channels at the ISDN port, also identifies the V5.2 link and time slots on the link that are to be allocated. This information element only identifies the mapping onto a single V5.2 link and does not allow existing connections to be overridden. Multiple-bearer-channel allocations to different V5.2 links require different ALLOCATION/Multi-slot-map messages, and allocations that override existing allocations require the single-bearer ALLOCATION/V5-time-slot-identification messages.

| | Bit 8 | Bit 7 | Bit 6 | Bit 5 | Bit 4 | Bit 3 | Bit 2 | Bit 1 | |
|---|---|---|---|---|---|---|---|---|---|
| Octet 4 | 0 | **Message type** | | | | | | | **Direction (AN:LE)** |
| | | **Allocation** | | ALLOCATION | | | | | <-- |
| | | | | ALLOCATION-REJECT | | | | | --> |
| | | | | ALLOCATION-COMPLETE | | | | | --> |
| | | **Deallocation** | | DEALLOCATION | | | | | <-- |
| | | | | DEALLOCATION-REJECT | | | | | --> |
| | | | | DEALLOCATION-COMPLETE | | | | | --> |
| | | **Audit** | | AUDIT | | | | | <-- |
| | | | | AUDIT-COMPLETE | | | | | --> |
| | | **Faults and Errors** | | AN-FAULT | | | | | --> |
| | | | | AN-FAULT-ACKNOWLEDGE | | | | | <-- |
| | | | | PROTOCOL-ERROR | | | | | --> |

**Figure 14.2** BCC protocol message types.

### 14.3.2 Allocation Complete and Allocation Reject Messages

The conclusion of a successful allocation process is indicated by the access network sending an ALLOCATION-COMPLETE message to the exchange. The only message-specific information element in this message is its BCC-reference-number, and this reference number is then free to be used by the exchange to

| | Message specific information elements | | | | |
|---|---|---|---|---|---|
| **Message** | A | B | C | D | E |
| Allocation | M | 1 | 2 | 3 | |
| Allocation-reject | | | | | M |
| Allocation-complete | | | | | |

Key:

A = User-port-identification B = ISDN-port-channel-identification C = V5-time slot-identification

D = Mutli-slot-map E = Reject-cause

**Figure 14.3** Message-specific information elements for allocation processes. Note: (M) Mandatory; (1) If the port supports more than one bearer channel; (2) If a single time slot is being allocated; and (3) If multiple time slots are being allocated.

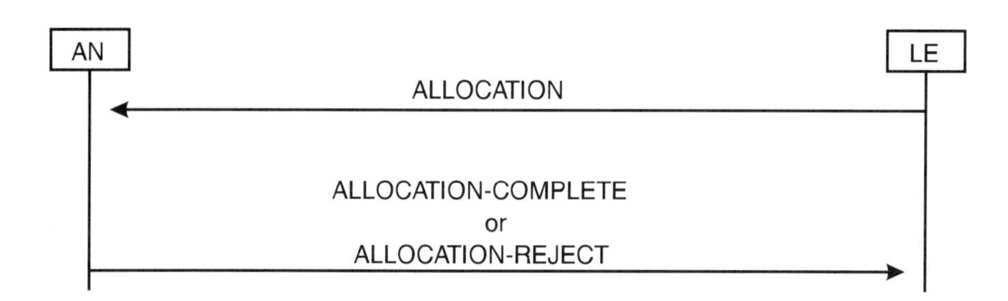


**Figure 14.4** Example message flow for bearer connection allocation.

identify another allocation or deallocation process. The access network also responds with an ALLOCATION-COMPLETE message if an existing allocation is requested once more.

If the access network cannot comply with an ALLOCATION message sent by the exchange, then it responds with an ALLOCATION-REJECT message that contains a Reject-cause information element. This information element contains a field that indicates the reason for the rejection, and for certain reasons this field is supplemented with a diagnostic field. Allocations can be rejected due to faults within the access network, at the user port, or at the V5 interface.

Allocations can also be rejected because of existing connections or even for no specified reason. The diagnostic fields contain information that may help to clarify the rejection. Allocation requests that attempt to override a connection that does not exist are also rejected, since these requests indicate that there is an inconsistency between the access network and its host exchange.

### 14.3.3 Deallocation Processes

The messages used for deallocation processes have formats and information elements that are identical with those used for allocation processes, since equivalent information is needed in both cases (see Figures 14.5 and 14.6). Normally a DEALLOCATION message marks the start of a new BCC process, but the exchange may also send a DEALLOCATION message to abort an allocation process using the process reference number of the initial allocation process. This does not cause confusion in the access network if the DEALLOCATION message is received after the access network thinks that the allocation process is complete, because it will then treat the deallocation as a new process which happens to have the same reference as the newly completed allocation, so the result will be no different.

| | Message Specific information Elements | | | | |
|---|---|---|---|---|---|
| **Message** | A | B | C | D | E |
| Deallocation | M | 1 | 2 | 3 | |
| Deallocation-reject | | | | | M |
| Deallocation-complete | | | | | |

Key:

A = User-port-identification B = ISDN-port-channel-identification C = V5-time slot-identification

D = Mutli-slot-map E = Reject-cause

**Figure 14.5** Message-specific information elements for deallocation processes. Note: (M) Mandatory; (1) If the port supports more than one bearer channel; (2) If a single time slot is being deallocated; and (3) If multiple time slots are being deallocated.

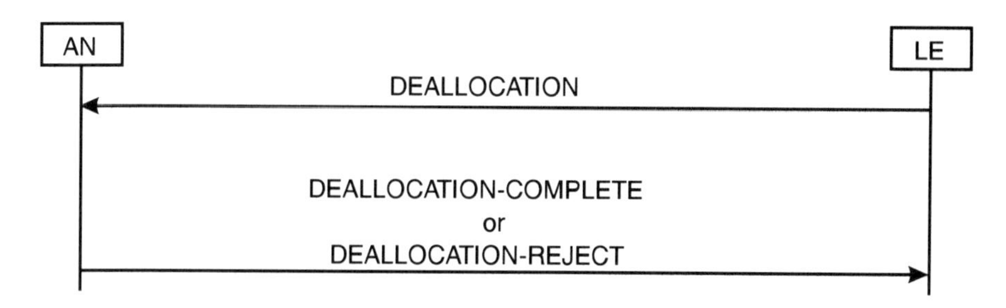


**Figure 14.6** Example message flow for bearer connection deallocation.

A successful deallocation is indicated by the access network sending a DEALLOCATION-COMPETE message. This message is sent even if there is no existing connection to deallocate, since in this case the deallocation process allows the exchange to confirm that connections have been cleared if there is confusion. A deallocation request can be rejected with a DEALLOCATION-REJECT message that has a Reject-cause information element, which may include additional values that are specific to deallocation processes and not used when rejecting an allocation.

### 14.3.4 Audit Processes

The purpose of an auditing process is to check the existing allocation of V5.2 bearer channels to bearer channels at the user ports. Auditing is initiated by the exchange, since it plays the role of the master in the V5.2 BCC protocol. Auditing can be performed either on a port basis or on a V5.2 basis, and the result indicates the port or V5.2 bearer channel that is complementary to the one identified in the request. In either case, auditing is performed on single connections, since this is the minimum necessary to meet requirements.

The exchange initiates an audit as a new process by sending an AUDIT message with information elements similar to a single-bearer-channel ALLOCATION or DEALLOCATION message, but which omits half of the information about the connection (see Figure 14.7). If a connection exists, then the access network responds with an AUDIT-COMPLETE message that contains the full information concerning the connection. If no connection exists, then the access network responds with an AUDIT-COMPLETE message with only two message-specific information elements: the BCC-reference-number, which identifies the original audit request from the exchange, and a Connection-incomplete information element to identify anomalous situations that may help resolve any confusion. There is no mechanism to reject an audit, since the result must either correspond to a specific channel mapping or to a incomplete connection.

| Message | Message Specific Information Elements | | | |
|---|---|---|---|---|
| | A | B | C | F |
| AUDIT | 4 | 5 | 6 | |
| AUDIT-COMPLETE | 7 | 8 | 7 | 9 |

KEY:

A = User-Port-Identification  B = ISDN-Port-Channel-Identification  C = V5-Time Slot-Identification

F = Connection-Incomplete

**Figure 14.7** Message-specific information elements for audit processes. Note: (4) If auditing by user port; (5) If auditing by user port with more than one bearer channel; (6) If auditing by V5 time slot; (7) If audit result is a complete connection; (8) If audit result is a complete connection and user port has more than one bearer channel; and (9) If audit result is that there is no complete connection.

### 14.3.5 Fault and Error Messages

The BCC processes for fault and error handling are the only BCC processes initiated by the access network. These processes are used to alert the exchange about faults in the access network that have an impact on the BCC protocol, or about BCC protocol errors that have been detected by the access network.

The access network sends an AN-FAULT message with a new BCC-reference-number to inform the exchange that an internal fault in the access network has affected a BCC connection. The format of the AN-FAULT message is similar to that of an ALLOCATE or a DEALLOCATE message for a single bearer channel, except that the information elements to identify the user port (and its channel for ISDN ports) and the V5.2 bearer channel are only included if they are known (see Figure 14.8). The exchange replies with an AN-FAULT-ACKNOWLEDGE message that has a mandatory BCC-reference-number as the only message-specific field with a value that indicates the initiating message.

If the access network detects a protocol error in a message received from the exchange, then it sends a PROTOCOL-ERROR message to the exchange with a new value in its BCC-reference-number information element. The only other information element in the message is the mandatory Protocol-error-cause, which identifies the type of protocol error and, where appropriate, the type of

| Message | Message Specific Information Elements | | | |
|---|---|---|---|---|
| | A | B | C | G |
| AN-FAULT | 10 | 11 | 10 | |
| AN-FAULT-ACKNOWLEDGE | | | | |
| PROTOCOL-ERROR | | | | M |

KEY:

A = User-Port-Identification B = ISDN-Port-Channel-Identification C = V5-Time Slot-Identification

G= Protocol-error-cause

**Figure 14.8** Message-specific information elements for faults and errors processes. Note: (M) Mandatory; (10) If known; and (11) If known and user port has more than one bearer channel.

message in which the error was detected and the erroneous information element in that message.

## 14.4 SUMMARY

The V5.2 interface allows bearer time slots on the interface to be allocated to specific user ports for the duration of an individual call. This dynamic allocation of bearer channels is controlled by the host exchange, because the access network is not always aware of when a call starts and finishes. On V5.1 interfaces, the allocation of bearer channels is static and not controlled by the interface.

The BCC protocol must be able to handle the allocation of bearer channels at the start of a call and the deallocation of the channels at the end of a call. It should also allow the allocation of channels to be checked and the access network to notify the exchange of faults that would prevent it from complying with a requested allocation.

Instead of always identifying the particular end points of the mappings, messages refer to the mappings by giving them a reference number. The exchange can send requests to the access network to allocate a specific V5 time slot to a specific channel at the user port, or to deallocate it. The access network responds with messages to indicate acceptance or rejection of these requests.

# 15 The Link Control Protocol

"The only thing that kept it standing was the woodworm holding hands."

—Jerry Dennis

An interface that consists of several parallel links generates additional requirements for the control of these links. These requirements, and the V5 messages that support them, are discussed in this chapter. The generalization to other interfaces is also considered.

## 15.1 CONTROL OF A MULTILINK INTERFACE

It should be possible to check that the two sides of an interface have been correctly interconnected by the physical link or links between them, since it is easy to become confused about the identity of links. If an interface consists of a single link, then identifying links can be achieved by giving each side of the interface a distinctive label. If several physical links have been disconnected, then they can be reconnected correctly by ensuring that the interface labels at either side match. This approach is more secure if the interface itself is used to confirm that the links match, because this also checks that each link is able to carry information.

This is the principle that is used for V5.1 interfaces. Either side of the interface is labeled by its interface ID and has the ability to check the interface ID at the other side. However, this approach is not adequate for V5.2 interfaces, because these interfaces have a number of links that correspond to the same interface ID label. To extend the principle of labeling to a multilink interface, it is necessary to label each of the individual links in addition to the complete interface, and a mechanism must be introduced to also check the link labels at each side of the interface.

In addition to the requirement to check link ID and integrity, it should be possible to take the links of a multilink interface in and out of service. This may be necessary either because a link has become faulty or because it is necessary to carry out planned maintenance. This requirement is similar to the requirement to block and unblock the ports on an access network. It differs from the procedure for ports, because it is possible to both protect signaling on a multilink interface and to minimize the disruption of bearer traffic. This difference has an impact on the need for blocking requests.

A blocking request allows the side that receives the request to minimize the disruption caused by blocking. For a port, disruption is minimized by allowing an ongoing call to be completed. For a link, there are two levels of reduction of disruption. A high-priority request allows disruption to be reduced by rapidly moving signaling traffic onto other links. A low-priority request also allows ongoing calls to be completed.

## 15.2 LINK INTEGRITY CHECKING

The identity and integrity of each link can be checked by selecting a particular link and sending a tracing signal over it. If the tracing signal arrives on the link with the label that matches the label at the transmitting side, then the link integrity check has been successful, since the two sides of the link match and the link itself can carry a tracing signal.

It should be possible to carry out this test in both directions, because these may be physically different connections. It is also necessary to avoid transmitting the same signal on more than one link at any time to avoid confusion. Ideally, different tracing signals should be used for different interfaces to avoid confusion if a number of interfaces are being tested simultaneously, but this has not been done for V5 interfaces, because the probability of accidentally returning a tracing signal on the correct link of a different interface is thought to be low.

The two sides of an interface must cooperate to agree on the logical label of the link that is to be physically tested with a tracing signal. This can be achieved if one side of the interface sends a message to request the other to return a tracing signal on a particular link. The side that receives the request can either comply with it or reject it, for instance, if it has no free tracing signal to apply to the link.

The tracing signal should not interfere with the normal operation of the link, because it may be necessary to check the link while it is in use. Since it may be desirable to be able to fill the link with bearer traffic, the tracing signal should make use of capacity in the link overhead. The tracing signal should also use very little bandwidth to maximize the bandwidth available to bearer traffic.

## 15.3 LINK BLOCKING AND UNBLOCKING

The functions that must be performed for the blocking and unblocking of links are very similar to those for the blocking and unblocking of ports. If a link fails or if it is not possible for management to wait any longer, then the other side of the interface must be informed that the link is no longer available.

In less dramatic situations, the access network must be able to request the exchange to block a link, because it is the exchange that is responsible for service and that has detailed knowledge of the ongoing traffic. The access network should be able to indicate the priority of the blocking request to make it clear whether or not it is prepared to wait for the completion of ongoing calls. The exchange need not request the access network to block a link, because the disruption caused by blocking a link has to be reduced by action taken by the exchange and not by the access network. However, the exchange must be able to inform the access network that a link is not available.

Unblocking of a link must be performed through cooperation at both sides of the interface unless the side that initiates blocking is kept informed of subsequent events at the other side of the interface.

## 15.4 V5.2 LINK CONTROL MESSAGES

Like all V5 messages, link control messages for the V5 interface consist of a generic header followed by message-specific information elements, with the first part of the header being the V5 protocol discriminator (see Figure 15.1). The message layer address in the header is used like that in the V5 control protocol (i.e., to identify what is being controlled). In this case it identifies the link to which the message refers.

The same, and equally redundant, message layer acknowledgments that are used for V5 control messages are also used for the link control protocol. The message type information element in the header identifies messages as either originating messages or as acknowledgments. This means that there are only two types of link control messages: LINK-CONTROL and LINK-CONTROL-ACK (see Figure 15.2). A LINK-CONTROL-ACK message is sent to acknowledge the receipt of the corresponding LINK-CONTROL message, but it only indicates receipt of the message and not acceptance of any requests.

The function associated with a message in the V5 link control protocol is identified by the single message-specific information element, the Link-control-function, which every link control message must contain. There are no other message-specific information elements for the messages of the V5 link control protocol. The functions specified by the values of the Link-control-function information element are either link identification or link blocking functions.

| | Bit 8 | Bit 7 | Bit 6 | Bit 5 | Bit 4 | Bit 3 | Bit 2 | Bit 1 |
|---|---|---|---|---|---|---|---|---|
| Octet 1 | Protocol discriminator | | | | | | | |
| Octet 2 | 0 | 0 | 0 | 0 | 0 | 0 | 0 | 0 |
| Octet 3 | Link Identifier | | | | | | | |
| Octet 4 | 0 | Message type | | | | | | |
| Octets 5, 6 &7 | Link-Control-Function | | | | | | | |

**Figure 15.1** Message format for the link control protocol.

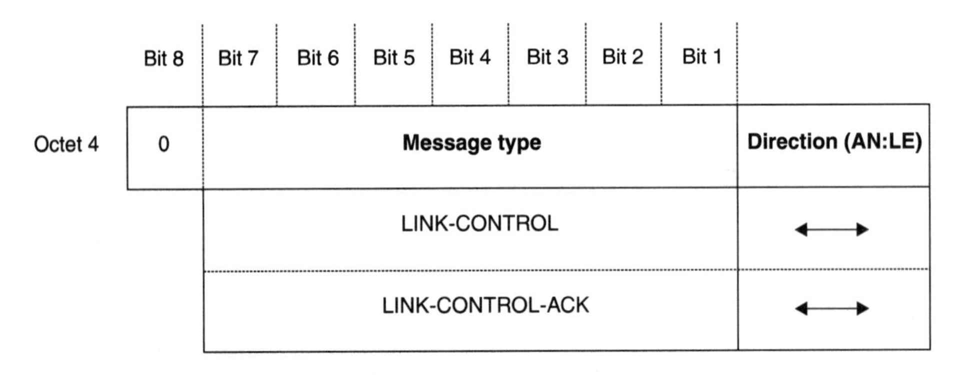


**Figure 15.2** Link control protocol message types.

### 15.4.1 V5.2 Link Identification Testing

The link identification procedure operates by one side of the interface requesting the other side to tag the link identified by the message layer address. Either side of the interface can initiate tagging by sending a LINK-CONTROL:link-identification-request message (see Figure 15.3). It can then check that the correct link has been tagged.

The side that receives the request can reject it. This would occur if, for instance, a previous link identification request were still being handled, because the specification of the tagging on the V5 interface does not allow more than one link to be tagged at any time. A request is rejected by sending a LINK-CONTROL:link-identification-rejection message (see Figure 15.4).

| Value of Link-control-function | Direction (AN:LE) |
|---|---|
| link-identification-request | ←→ |
| link-indentification-rejection | ←→ |
| link-identification-acknowledgment | ←→ |
| link-identification-release | ←→ |

**Figure 15.3** Forms of the link control message for link identification testing.

If the request is accepted by the receiving side, then it tags the indicated link and responds by issuing a LINK-CONTROL:link-identification-acknowledgment message to confirm that this has been done. A better name for this would be link-identification-compliance, because this would avoid possible confusion with the V5 LINK-CONTROL-ACK messages. Tagging is applied to a link by setting the Sa7 bit of the link to 0. It is removed by returning the Sa7 bit to 1.

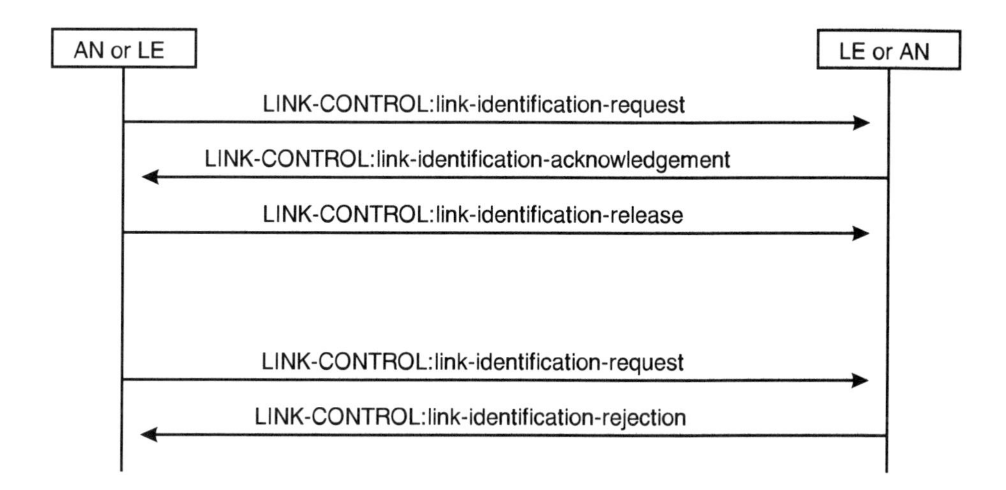


**Figure 15.4** Example message flows for link identification testing.

When the side that issues the request has received the indication of compliance and has checked the tagging, it may request the removal of tagging by sending a LINK-CONTROL:link-identification-release message. When the other side receives the release message, it removes the tagging.

The added value of the LINK-CONTROL-ACK messages, which are sent on receipt of the associated LINK-CONTROL messages is questionable. These messages are not needed to ensure the integrity of the communications across the interface, because the frame-layer of the protocol performs this function. Eliminating these messages from the link identification procedure would mean that there would be no message to acknowledge the request to stop the tagging, but this request is implicitly acknowledged when the tagging itself is stopped. For these reasons, it is not clear that the LINK-CONTROL-ACK messages serve any useful purpose for the V5 link identification procedure.

### 15.4.2 V5.2 Link Blocking and Unblocking Messages

The messages that are used to block and unblock links in the link control protocol are similar to the messages used to block and unblock ports in the control protocol. The main difference is that there are two types of link blocking requests that can be issued by the access network (see Figure 15.5).

The two link blocking requests indicate different levels of urgency, a distinction that is not made for port blocking requests. This distinction is required, because the host exchange may reject a blocking request. If this occurs for a port, then the access network may insist by blocking the port unilaterally.

| Value of Link-control-function | Direction (AN:LE) |
|---|---|
| deferred-link-blocking-request | ⟶ |
| non-deferred-link-blocking-request | ⟶ |
| link-block | ⟷ |
| link-unblock | ⟷ |

**Figure 15.5** Forms of the link control message for link blocking.

However, if the access network unilaterally blocks a link, then this is likely to cause much greater disruption than if it unilaterally blocks a port. To avoid this, it is useful for the access network to have a way to indicate how urgently the link should be blocked.

The access network can request the exchange to block the link by sending it a LINK-CONTROL:deferred-link-blocking-request message or by sending a LINK-CONTROL:non-deferred-link-blocking-request message (see Figure 15.6). The LINK-CONTROL:deferred-link-blocking-request message is less urgent, since it indicates that the access network is prepared to wait for the exchange to switch any communications channels onto a different link and to wait for any ongoing calls to be completed.

The LINK-CONTROL:non-deferred-blocking-request is more urgent, since it indicates that the access network is not prepared to wait for ongoing calls to be completed. However, it does indicate that the access network is prepared to wait until the exchange has switched any communications channels onto a different link. The nondeferred blocking request can be used if the access network has become impatient after deferred blocking has been requested and blocking has not been completed. A LINK-CONTROL:link-block message could be used instead of the LINK-CONTROL:non-deferred-link-blocking-request if there are no communications channels on the link, but this is not recommended because it is safer to give the exchange some warning of the intention before indicating that the link is not available.

The host exchange does not need to request the access network to block a link, because the exchange is aware if calls are in progress and can control the use of time slots on the V5 interface. If the exchange wants to block a link, it can use the protection protocol to switch communications channels onto time slots on a different link, and it can use the BCC protocol to ensure that no more calls use the bearer time slots. After all ongoing calls have terminated, the

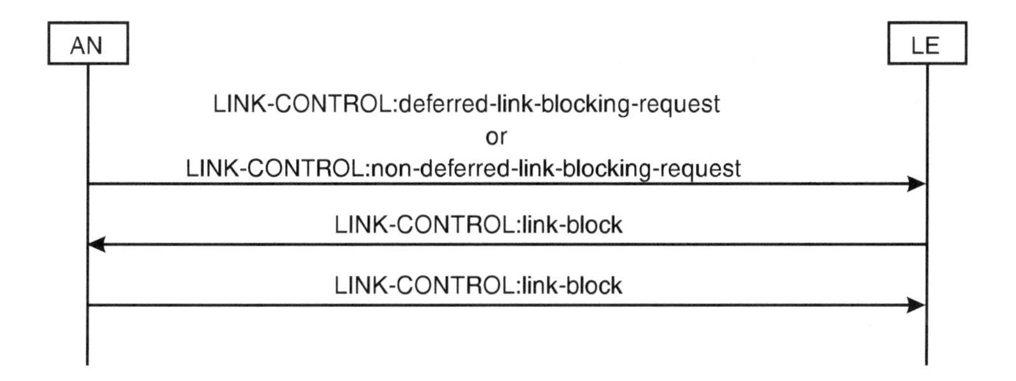


**Figure 15.6** Example message flows for link blocking requests.

exchange can send a LINK-CONTROL:link-block to inform the access network that the link is blocked.

The V5 link control protocol does not include the possibility of the exchange sending blocking requests to the access network, because these are unnecessary. The access network cannot assist the exchange with link blocking, because the exchange already has both the capability and all the necessary information.

The access network can also spontaneously send a LINK-CONTROL:link-block message to inform the exchange that the link is blocked, but this is likely to disrupt service, because the link may be in use. The access network may still do this, for instance, if there is a failure on the link or if previous requests for the exchange to block the link have been ignored.

The access network and the exchange must cooperate to unblock a previously blocked link, because the access network and the exchange are autonomous, and so each may independently undertake maintenance after the other has blocked the link and the V5 protocols do not include the possibility of each side informing the other about this. The unblocking works in the same way as the unblocking of ports in the V5 control protocol. When either side wants to unblock a link, it sends a LINK-CONTROL:link-unblock message to the other side. If the other side agrees to unblock the link, it replies with a LINK-CONTROL:link-unblock message of its own. Both sides must have transmitted and received a LINK-CONTROL:link-unblock message before the link is unblocked.

The LINK-CONTROL-ACK messages have some use when unblocking links, because they give confirmation that the LINK-CONTROL:link-unblock messages have been received. This means that both sides of the link can move to an unblocked state, confident that the other side has received their unblocking message and knowing that they have received an unblock message from the other side. They also confirm that a LINK-CONTROL:link-block message has been received so that the side that has sent the message can be more comfortable about proceeding with maintenance on the link.

This usefulness of the LINK-CONTROL-ACK messages is limited, because the frame layer of the protocol is responsible for ensuring that messages are delivered correctly. The message layer should always be free to assume that messages are delivered correctly and should be able to rely on the frame layer to tell it otherwise. Any problem delivering messages indicates a serious fault with the interface, which should be handled accordingly. The link control protocol is not responsible for the correct operation of the frame layer.

It is difficult to see what the usefulness of the LINK-CONTROL-ACK messages is for the link blocking requests, because the desired response to a link blocking request is a LINK-CONTROL:link-block message, not an acknowledgment. It is not clear that the cost of implementing these messages will be offset by corresponding advantages.

## 15.5 OTHER INTERFACES

The V5.2 link control protocol has been invented to handle multiple physical links because the V5.2 interface is not restricted to a single physical link. The principles involved are also applicable to other interfaces with several physical links, because the same considerations of link maintenance and integrity checking apply. The details may differ, since these may depend on the physical characteristics of the links.

It is reasonable to assume that it is possible for blocking and unblocking of other physical links to be performed in the same way as for V5 interfaces. The reason for this is that the V5 protocols demonstrate that essentially the same blocking and unblocking procedure can be used for different physical interfaces, namely, ISDN ports, PSTN ports, and V5.2 links, and so the approach appears to be independent of the physical characteristics. However, it is not so clear that the V5 mechanism for blocking and unblocking is the best approach for other interfaces because, it is not consistent with ITU-T Recommendation X.731 on state management.

The considerations that led to the invention of the link identification procedure for the V5 interface also apply to other interfaces with several physical links. However, the particular details of link identification may differ because the V5 protocol is constrained by the nature of the physical links on the V5 interface. A simple, single tagging signal is used to identify V5 links, because it is easy to include this in the 2.048-Mbps structure. Other physical links would allow other tagging signals to be used. This could allow several links to be tagged, or allow every link to be continuously labeled with a unique tagging signal. The latter possibility would eliminate the need to request a tagging signal to be applied, because tagging would be present continuously. This could have been adopted for the V5 interface if a more complex tagging signal had been used. So although the same considerations make it desirable to implement link integrity checking for other interfaces, the implementation of the function could be quite different than on the V5 interface.

Other interfaces may not require as many links as the V5 interface, especially if they operate at higher rates. A single STM-1 bearer has four times the capacity of the largest V5.2 interface and only requires a single physical link in each direction. It would be possible to multiplex a number of V5 interfaces onto a single physical link, so that each interface consisted of a number of logical links on a shared physical link. It would also be possible for an interface to consist of virtual links on an ATM network.

At first sight there appears to be no need for a link control protocol when there is a shared physical link, because the control of the physical link is not the responsibility of the control of the logical or virtual interfaces that it supports. This view is simplistic, because there is still a need for the control of logical or virtual links of the interface. The need for blocking and unblocking of nonphysical or logical links may be even higher than for physical links because blocking of the shared physical link would affect all the traffic.

There may also be a higher need for integrity checking of nonphysical links, because a physical link normally requires the physical presence of a human being to disrupt it. Nonphysical links can be disrupted far more efficiently, because this can be done remotely by a human operator or even by a software bug. It can also be done less obviously, since the change cannot be detected at the physical layer.

In conclusion, the considerations that lead to the link control protocol of the V5 interface also apply to interfaces with several links, regardless of whether the links are different physical links, logical links, or virtual links. The requirements for link blocking and unblocking and for link identification may be even greater because of the need to manage additional traffic on the same physical bearer and because it may be easier for link integrity to be accidentally violated. The procedure to block and unblock links appears to be general, because it is already applied to different physical interfaces through the V5 protocol, but it may need to be modified, because it is not consistent with CCITT Recommendation X.731. The procedure to check link identification will probably differ, because it depends on the physical nature of the interface, and the complexity of the procedure can be traded off against the complexity of the tagging.

## 15.6 SUMMARY

Interfaces with more than one physical link generate additional requirements for their control. In particular, it is necessary to be able to block and unblock the links so that the interface can be maintained, and it is necessary to be able to check the physical configuration of the links to check that they have been correctly interconnected.

The V5 link control protocol allows either side of the interface to request the opposite side to tag the link identified in the message address field. This request can be accepted or rejected. If it is accepted, then the tagging continues until released by the side that has requested it. A request may be rejected if a link is already tagged, because the implementation of this function on a V5 interface only allows one link to be tagged at a time.

The blocking and unblocking of the links on the V5 interface is similar to the blocking and unblocking of user ports on the V5 control protocol. It is slightly more sophisticated, because there are two priorities of blocking requests for links. A low-priority request allows the host exchange both to switch communications channels on to other links and to wait till ongoing calls are completed. A high-priority request only allows the communications channels to be switched, and may result in the loss of ongoing calls.

The V5 link control messages have no optional information elements and only one message-specific information element that identifies the link control function. The link to which the message refers is identified in the message layer address field.

The link control protocol could be generalized to other applications. It could control physical links at other multiplexed rates, or it could control virtual or logical links. A different application need not be constrained to the tagging of a single link at a time, and it could use different tagging signals for different interfaces or links, but the same principles apply and similar messages could be used. Other applications could benefit from simplification of the relationship with CCITT Recommendation X.731.

# 16 The Protection Protocol

"It works better if you plug it in."
—Arthur Block (Sattinger's Law)

The V5.2 protection protocol does not contribute directly to the functioning of the interface because it is the other protocols that are responsible for the primary functionality. The contribution the protection protocol makes is indirect, by ensuring that the other protocols can still operate after an equipment failure or if a link is accidentally unplugged. The protection protocol also assists in moving active communications channels off a link that is about to go out of service.

In the specification of the V5 interface, it has been assumed that the most likely type of equipment failure at the V5.2 interface is the failure of one of its links. It is also assumed that it is sufficient to protect against the failure of a single link. As more experience is gained with implementations, it will be possible to check the validity of these assumptions. In particular, they may be optimistic because additional links way be susceptible to the same factor that causes a single link to fail.

The principles and architecture for protection switching on a V5 interface are described first, followed by a description of the messages used in the protocol. Because the concept of protection is not limited to the V5 interface, the form that a protection protocol might take on other interfaces is also examined.

## 16.1 THE PURPOSE OF THE PROTECTION PROTOCOL

The purpose of the protection protocol is to ensure that the other protocols can still operate despite an equipment failure so that the operation of the V5.2 protocol is more robust. In particular, it is meant to safeguard the logical communications channels against the failure of a single link on the V5.2 interface.

### 16.1.1 Primary Functionality

Bearer channels, unlike communication channels, are implicitly protected if they are dynamically allocated to user ports. A connection may be interrupted if a bearer channel on the interface fails, but it can be rerouted to another bearer channel if bearer channels are flexibly allocated. A similar mechanism is required for communications channels to protect them against equipment failure. The protection of communications channels is more important than the protection of bearer channels, because the failure of a communications channel affects a number of user ports. This is especially clear if the communications channel that supports the protocol for the allocation of bearer channels fails, since then the implicit protection of bearer channels is lost.

To protect the communications channels, it is helpful to distinguish between the secure logical communications channels and the insecure physical time slots with which they are associated. When a failure affects the physical time slot, then the logical channel must be reallocated to a different physical time slot. This mapping between logical channels and physical time slots is similar to the mapping between channels at the user ports and bearer time slots on the V5.2 interface.

The protocol that protects communications channels must be split between links because this protocol must also have a means to protect itself; otherwise there could be no protection against the failure of the link that supported the protection protocol. The protection protocol for communication channels cannot rely on yet another protocol for its own protection because this would lead to an infinite regression.

There are two categories of protected communications channels. There is the logical communications channel used by the protection protocol, and there are the other channels, which are protected by the protocol. Because protocols may share communications channels, this means that there are two ways of protecting protocols. Protocols that share the same logical communications channel as the protection protocol are automatically protected by the same mechanism the protection protocol uses to protect itself. Other protocols are protected if they use the logical channels that are protected by the protection protocol.

There must be no confusion or conflict between the two sides of the interface concerning the mapping of logical communications channels onto physical time slots, since both sides of the interface have to be able to identify the logical channels correctly. The simplest way to ensure this is for the mapping to be controlled by one side of the interface, and for the other side of the interface to issue requests and be informed of changes. Both sides of the interface should be able to initiate the switch to a different physical time slot, because a failure might only be detected on one side.

It is sensible for the exchange to control the mapping between logical communications channels and physical time slots, because this mapping affects services for which the exchange is responsible.

### 16.1.2 Secondary Functionality

If protocol errors are detected on the protection protocol, then they should be flagged, because problems here may have serious consequences for the interface. In particular, the controlling side of the interface should be informed if protocol errors are detected at the other side, because these errors may require the mapping from logical channels to physical time slots to be changed.

It is also possible for the protection protocol on either side of the interface to become out of step, for example, if messages are lost. If this happens, then it may be necessary to reset the protocol. Because the protection protocol affects the operation of all user ports on the interface, it may be sensible to introduce additional handshaking to ensure that both sides are fully aware that a reset is occurring.

## 16.2 FUNDAMENTALS OF V5.2 PROTECTION SWITCHING

The communications protocols of a V5 interface are carried on communications paths that are statistically multiplexed onto logical communications channels that have a bandwidth of 64 Kbps. The communications protocols are protected by protecting these logical channels rather than protecting the individual communications paths. This approach was chosen because it is simpler to do circuit switching of the multiplexed 64-Kbps channels than to do packet switching of every communications path.

The messages of the protection protocol itself are transmitted twice, once on each of the two links that carry it. This is unlike any other V5 protocol. The two links involved use time slot 16 for the protection protocol, because this is compatible with the use of time slot 16 for the control protocol on a V5.1 interface. Two links are used to ensure that the protection protocol is intrinsically protected against the failure of a single link and does not need a second protection protocol, which would need a third, and so on ad infinitum. This approach also confirms that the two time slots that carry the protection protocol are fully functional. These two time slots form Protection Group 1.

The other V5 protocols that are not used for call control also use the time slots of Protection Group 1, but unlike the protection protocol, the messages of these other protocols are not simultaneously transmitted on both time slots. The link used by these protocols on initiation is known as the *primary* link of the

V5.2 interface. The other link used by the protection protocol is known as the *secondary* link. The protocols that initially use the primary link are switched onto time slot 16 of the secondary link by the protection protocol if the primary link fails.

It makes sense to use the time slots of Protection Group 1 for other housekeeping protocols, because the protection protocol only uses a small fraction of the bandwidth of a 64-Kbps time slot. These housekeeping protocols are more important than the call control signaling because all user ports may be affected if they fail, whereas the failure of one of the communications paths for call control only affects some of the user ports. Using Protection Group 1 for housekeeping protocols ensures that there is always a standby time slot reserved to protect them.

Call control protocols may also be protected using Protection Group 1, but typically there will not be sufficient bandwidth remaining to do this. Other physical time slots may be allocated to another protection group, Protection Group 2, which can only be used for call control protocols. The logical communications channels belonging to Protection Group 2 are allocated to certain of these physical time slots and the remaining physical time slots remain in standby. Frame layer flags are constantly transmitted in the standby time slots of Protection Group 2 so that the error rate on the standby channels can be monitored.

The V5 specification does not insist that Protection Group 2 is used, because it may be too elaborate for certain simple applications. Typically, there will be fewer standby channels in Protection Group 2 than logical communications channels, because it is not necessary to have more than three standby channels to protect against the loss of a single link. Unlike Protection Group 1, Protection Group 2 was not designed to provide each active time slot with its own standby time slot. Instead, the standby time slots provide a shared resource that protects the active time slots.

A protection switch of the logical communications channel onto a different physical time slot is initiated using the protection protocol if excessive errors are detected on a protected communications channel. These errors may be detected at the frame layer or at other layers. Switching of logical communications channels onto different physical time slots may also be initiated for other reasons, for instance, because it is necessary to carry out maintenance or to block a link.

## 16.3 V5.2 PROTECTION PROTOCOL MESSAGES

The generic header of the messages of the V5.2 protection protocol begins with the protocol discriminator common to all V5 messages (see Figure 16.1). It ends with the message type information element that identifies the message as one

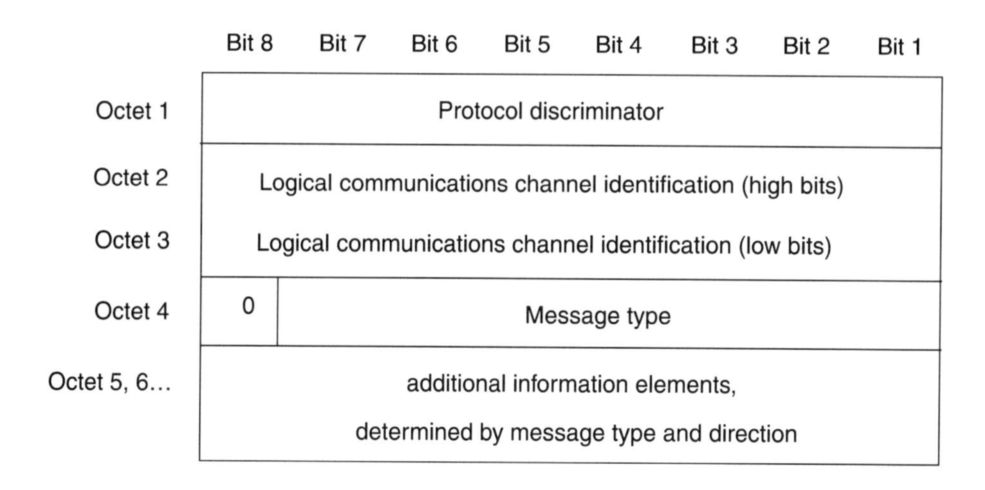


**Figure 16.1** Protection protocol message format.

of the eight possible protection protocol messages. The message layer address, which is between the protocol discriminator and the message type, is used for protection protocol messages to identify the logical communications channel to which the message refers.

The message-specific information elements following the generic header are always mandatory and are determined by the message type. Five of the eight messages, the switch-over messages, control the association of the logical communications channels with physical time slots (see Figure 16.2). Protocol error messages are sent to indicate errors in the format of received messages. Switch-over messages and protocol error messages are numbered in sequence. Reset messages are sent as commands or acknowledgments if problems with the sequence numbering of the other messages are discovered.

### 16.3.1 Switch-Over Messages

The V5.2 protection protocol defines five separate types of switch-over messages. These messages are numbered in sequence using the Sequence-number information element (see Figure 16.3). The physical time slot to which these messages refer is identified by their Physical-C-channel information element. All switch-over messages must include these information elements, and the SWITCH-OVER-REJECT messages must also include a Rejection-cause information element, which gives the reason for rejecting a switch-over.

The commands that switch the logical communications channels onto different physical time slots are issued only by the host exchange (see Figure 16.4). If the switch-over has been initiated by the operations system of the host

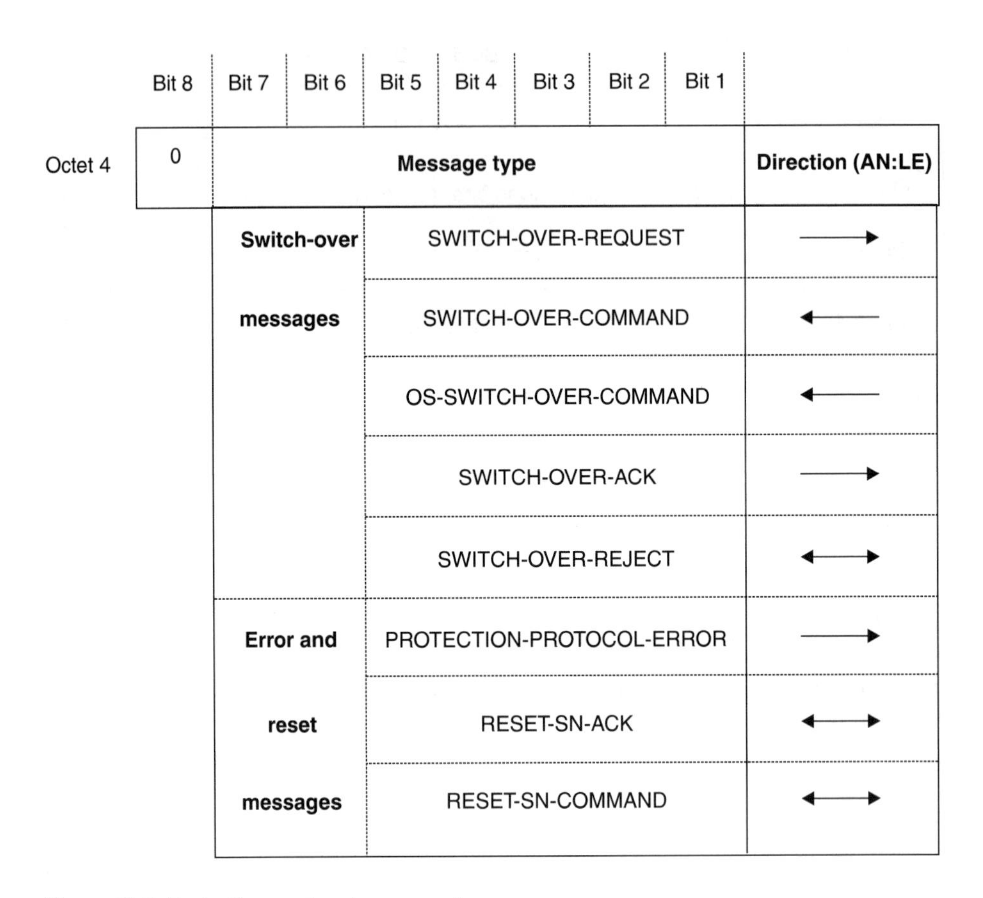


**Figure 16.2** Protection protocol message types.

exchange, then the exchange sends an OS-SWITCH-OVER-COM message to command the access network to switch the identified logical communications channel onto the identified physical time slot. The exchange can also send a SWITCH-OVER-COM message to perform the same function if it is not appropriate to specify that the switch-over has been initiated by the operations system of the host exchange. However, there appears to be no clear reason for being able to inform the access network whether or not the operations system of the host exchange initiated the switch-over.

The access network sends a SWITCH-OVER-ACK message to inform the host exchange that it has accepted the command to switch the logical communications channel to the new physical time slot. If the access network is not prepared to accept the command from the host exchange, then it replies with a SWITCH-OVER-REJECT message.

| Message | Message Specific Information Elements | | |
|---|---|---|---|
| | A | B | C |
| SWITCH-OVER-REQUEST | M | M | |
| SWITCH-OVER-COMMAND | M | M | |
| OS-SWITCH-OVER-COMMAND | M | M | |
| SWITCH-OVER-ACK | M | M | |
| SWITCH-OVER-REJECT | M | M | M |

KEY:

A = Sequence-number    B = Physical-C-channel-identification    C = Rejection-cause

**Figure 16.3** Message-specific information elements for switch-over messages. Note: (M) Mandatory.

The access network can also use a SWITCH-OVER-REQ message to request the host exchange to switch an identified logical communications channel onto an identified physical time slot. The host exchange holds the master table for the mapping of the logical to physical communications, because this mapping

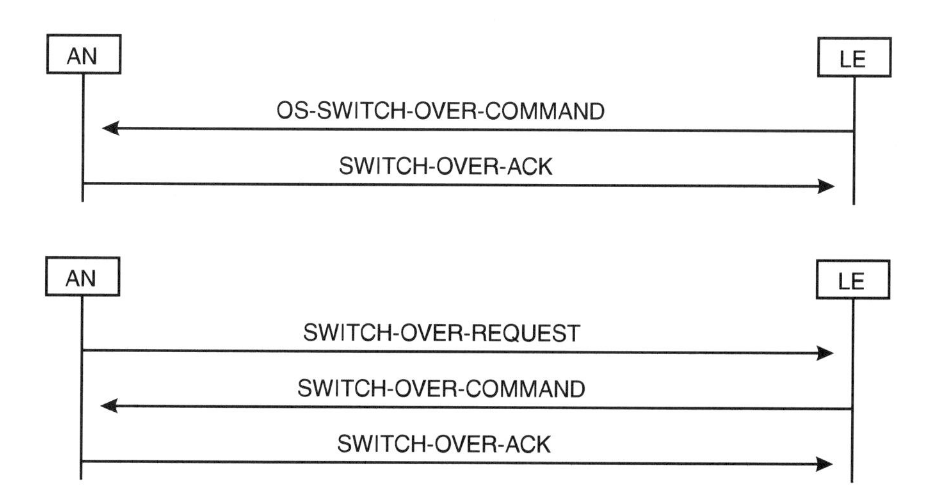


**Figure 16.4** Example message flows for switch-over.

affects the ability to support services. The access network can only make requests and proposals because it does not control the mapping.

The exchange can reject a switch-over proposed by the access network by using a SWITCH-OVER-REJECT message that also identifies the problem. Switch-over reject messages are the only switch-over messages that can be sent by either side of the interface. Switch-over commands are only sent by the exchange and switch-over requests and acknowledgments are only sent by the access network.

### 16.3.2 Reset Messages

The reset messages of the V5.2 protection protocol are used to reset the sequence numbering of the switch-over and protocol error messages if a fault is detected. Each side of the interface expects to receive messages that are consecutively numbered, and a fault is registered if there is a jump in this numbering.

Reset messages contain no message-specific information elements, and so consist only of a generic header. The side that detects the loss of alignment sends a RESET-SN-COM message to inform the other side that it has reset the sequence numbering. The side that receives a RESET-SN-COM message replies with a RESET-SN-ACK message to acknowledge that it is aware that the sequence numbering has been reset.

Reset messages are not specific to individual logical communications channels. There is no significance attached to the message layer address in the generic header, and this is set to 0.

### 16.3.3 Protection Protocol Error Message

The V5.2 protection protocol contains a single type of protocol error message, which is used by the access network to inform the exchange that it has detected a protocol error. The PROTOCOL-ERROR messages contain a Protocol-error-cause information element, which indicates the type of error and may also indicate which part of the received message is at fault (see Figure 16.5).

Like the various types of switch-over messages, PROTOCOL-ERROR messages are numbered in sequence by their Sequence-number information element. Unlike the switch-over messages, they do not refer to any physical time slot, only the logical communications channel identified in the generic header.

## 16.4 GENERALIZATION TO OTHER INTERFACES

The considerations that apply to the V5.2 protection protocol are also relevant for other interfaces, since other interfaces may also need to protect their com-

| Message | Message Specific Information Elements |
|---|---|
| PROTECTION-PROTOCOL-ERROR | Sequence-number and Protocol-error-cause (both mandatory) |

**Figure 16.5** Message-specific information elements for the protection protocol error messages.

munications protocols against equipment failure or to move active communications channels between links. Certain of the details of the V5.2 protection protocol could have been specified differently without altering the basic functionality of the protocol. Other details are specific to the V5 architecture and the messages would have to be given a more general interpretation if applied to another interface.

The principle that communications should be protected against the failure of a single V5 link is specific to the V5 interface because other interfaces may need to be protected against other types of failure. Even in the V5 case, it might be more sensible to protect against multiple link failure, because an event that causes a single link to fail could result in the failure of several links. Other interfaces could use logical or virtual links, so they may require protection within a single physical link.

In general, the V5.2 protection protocol can be interpreted as controlling the association of protected entities with the resources that support them. The protected entities could be the communications protocols of the other interfaces. This interpretation allows the protection to be applied on single physical links, or on the basis of communications paths rather than communications channels.

If the protocol is generalized in this way, then certain of the details of the V5 protocol would have to be changed, because they are specific to the V5 architecture. The detailed coding of the information elements reflects the nature of the V5 interface. Another V5-specific detail is the existence of two switch-over commands. This is not a general requirement, because the functions performed by local management and by an operations system vary from implementation to implementation.

The V5-specific principle that the communications should be protected against the failure of a specific link is ingrained in the use of two links for the protection protocol, because the interface would fail if both of these links fail. A greater level of protection could be achieved if the protection protocol were not constrained to just these two links. The protection protocol could use any physical channel that was capable of transmitting communications, because the channel used could be identified automatically. This identification could be achieved by examining frame layer addresses to check for those corresponding to the protection protocol, or by appropriate tagging of the correct link.

Flexibility, both on the allocation of other communications protocols to physical resources by the protection protocol and on the physical resources used by the protection protocol, is the key to the survival of the interface in the event of physical failure. In order to survive an equipment failure, the interface should be capable of operation on whatever the remaining resources are. Tagging may be useful here, because it could be used to indicate the location of the protection protocol on the surviving resources, and this could then be used to bootstrap the interface back into operation.

There is an analogy between a protection protocol and the use of error correcting codes. In both cases, the correct interpretation must be made despite the presence of faults. This is achieved through including redundancy which allows the same information to be conveyed a number of different ways, and the raw data can be processed to provide the correct interpretation. In one case this is the correct bit pattern, and in the other it is the correct configuration for communications channels.

A different interface might not be able to define Protection Group 1, because it might not intrinsically protect the protection protocol by simultaneously transmitting on two time slots, which are then identified as the active and standby for Protection Group 1. A different approach to the intrinsic protection of the protection protocol could also eliminate the need for sequence numbering and for the associated reset messages, because this is driven in part by the need to identify messages sent on the two channels.

The PROTOCOL-ERROR message could be moved from the protection protocol to the general part of the Control protocol, because it is really a general function that is not specific to the protection protocol. If this were done in addition to removing the reset messages and only having a single switch-over command, then the protection protocol could be greatly simplified. Whether the PROTECTION-ERROR message is moved or not, it could be modified so that it could also be sent by the exchange, because the access network should be informed if the exchange is aware of protocol errors in the messages it receives, since these errors could be due to a fault or bug in the access network.

## 16.5 SUMMARY

The purpose of the V5.2 protection protocol is to ensure that communications channels are protected against equipment failures, particularly the failure of a single link. This is achieved by switching the logical communications channels onto different physical time slots if a failure occurs. The protection protocol itself cannot be confined to a single physical link, because it would then be vulnerable to the failure of the link that transported it.

The protection protocol is carried on time slot 16 of both the primary and the secondary V5.2 links, because this gives greater security. To avoid wasting bandwidth, the housekeeping protocols also share these time slots. These are normally carried on time slot 16 of the primary link, but are switched to time slot 16 of the secondary link by the protection protocol if the primary link fails. These two time slots form Protection Group 1.

Call control protocols may also be carried on the same logical channel as the housekeeping protocols, but this may be difficult because of the bandwidth required. If other logical channels are used, then they are switched between the physical time slots of Protection Group 2 by the protection protocol. Unlike the standby channel of Protection Group 1, which is constantly monitored by the transmission of protection protocol messages to confirm that they are fully functional, the standby channels of Protection Group 2 are only partially monitored, since no messages are sent over them until a switch-over occurs.

The commands that switch the logical communications channels onto different physical time slots are issued only by the host exchange. The access network can only request a switch-over and acknowledge commands from the exchange. Either side can reject a switch-over, but the message to do so must indicate the reason for the rejection. Rejections, like all other switch-over messages, must also identify a physical communications time slot. All switch-over messages are also numbered so that it is possible to identify the two copies of the messages that are sent over the two physical time slots used by the protection protocol.

Protocol error messages are also numbered for the same reason, and like switch-over rejection messages, they must indicate the origin of the problem, but unlike switch-over rejection messages, they do not identify a physical time slot. If the numbering of messages becomes misaligned, then a message can be sent to command the resetting of the numbering, and an acknowledgment of this can be returned.

The principles of protection switching used on the V5 interface could also be applied to other interfaces where protection against failure is required. The need for the protection protocol to be automatically immune to failure is not specific to the V5 interface, but other interfaces could use very different approaches. The association of the other housekeeping protocols with the protection protocol could be altered because other approaches may not use two time slots for the protection protocol. This would also affect the sequence numbering and need for messages to reset this. On other interfaces, communications could be protected according to the communications protocol, rather than by a communications channel that several protocols may share. The V5 protection protocol should perhaps be thought of as an example for analysis and refinement for other interfaces, rather than an approach to be copied without detailed re-examination.

# 17 The VB5 Interface

"Fog in the English Channel. The continent is isolated."
—Anonymous

After the successful development of the specifications for the narrowband V5 interfaces, it was natural for standards organizations to consider developing the corresponding broadband interface. Following an initial report, ETSI started work on this in earnest in 1995 and the interface became known as the VB5 interface. The story of the work since then is characterized by a succession of changes of direction.

Many of those organizations involved in the work showed a strong preference for a minimalist approach, which ideally would have avoided the need to specify any new protocol. The hope of the minimalists was that the OAM flows that had already been defined for ATM would provide sufficient functionality to avoid the need to specify blocking messages for VB5. This is not realistic, since an examination of the V5 interface reveals the need for blocking request messages, which cannot be handled by the OAM flows previously defined. One of the justifications for the minimalist view was speed of implementation, which is ironic, since discussions about it may have delayed implementations more than if the minimalist view had been discarded immediately. However, there was an agreement that, unlike the V5 interface, the VB5 interface would be directly compatible with ITU-T Recommendation X.731, or so it seemed.

Much of the discussion at this point focused on architectural considerations, and the distinction was then drawn between VB5.1 interfaces, for broadband access networks performing only VP cross-connection (see Figure 17.1), and VB5.2 interfaces, for broadband access networks that could also perform VC switching (see Figure 17.2). This VC switching could be controlled by the responsible associated ATM switch, the service node, through the broadband generalization of the V5 BCC protocol. Unfortunately, this distinction between VB5.1 and VB5.2 was not to endure. It did, however, demonstrate a further

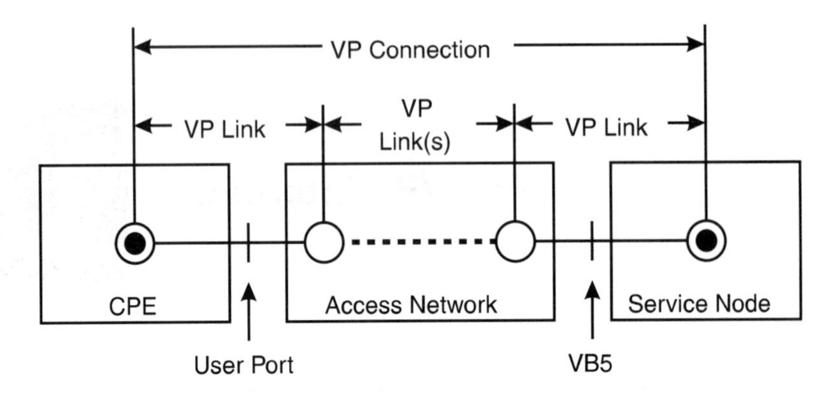


**Figure 17.1** VB5 with VP cross-connection.

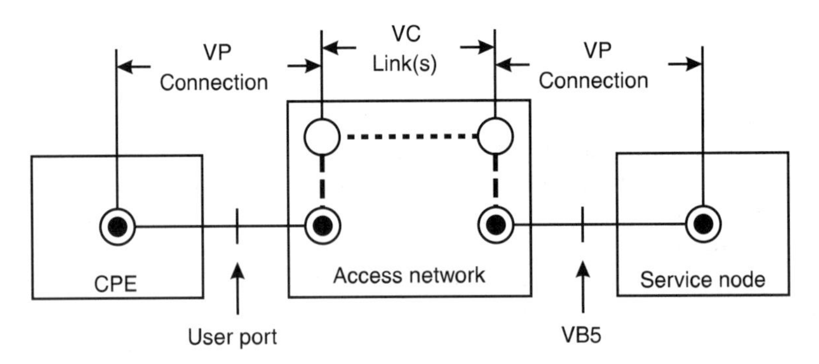


**Figure 17.2** VB5 with VC switching.

problem with relying on the defined OAM flows as a means of conveying fault indications. It was unfortunate enough in the VB5.1 case for a fault at an ATM user port to generate an F4 OAM flow for each of the VPs at the port. However, in the VB5.2 case, these VPs terminated in the access network, causing a further cascade of F5 OAM flows that could flood the ATM switch, all due to a single fault at a user port.

The original distinction between VB5.1 and VB5.2 interfaces died because it was realized that VC switching was too much of a good thing for the VB5.2 interface to keep to itself. It was decided that VC cross-connection could be performed in an access network with a VB5.1 interface, but VC switching controlled by the ATM switch could only be performed in an access network with a VB5.2 interface. At this point, the minimalists were having a bad time. Having been forced to concede that something like the V5 control protocol was required to handle blocking requests, they now found the use of OAM flows to convey fault indications under threat, because it would cause the ATM switch to be flooded, even for the VB5.1 case if there were VC cross-connections in the access network.

However, it was the agreement on ITU-T Recommendation X.731 that led to the final massacre of the minimalists. There had been an assumption that the OAM flows could be used as fault indications, while messages could be used for block requests. Unfortunately, according to the interpretation of X.731 for ATM, an administrative action could interfere with the OAM flows, preventing them from being available to carry the necessary fault indications. It became clear that additional messages would be needed for fault indications, since the OAM flows alone were not sufficient.

Unfortunately, ITU-T Recommendation X.731 proved to be a double-edged sword. An administrative action could also affect dependent entities in the same way as a real fault. This caused difficulty for the definition of the new messages, since it was intended to keep a clear distinction between real faults and the effects of administrative actions. Since something had to give here, it was decided that where it was possible, the VB5 messages would clearly differentiate between faults and the effects of administrative actions, even if this distinction were not so clear in other existing standards.

However, just as the crows were descending on the corpses, it became clear that the real battle had not even begun. So far the debate had been between members of ETSI and had something of the nature of a family quarrel. Participants at ITU-T and the ATM Forum now realized that the VB5 interface was far too important to leave in the hands of continental Europeans.

The view was expressed within ITU-T that these management coordination messages were really between blocks of operations systems functionality physically implemented within TMN network elements, and so the communication should not involve a new protocol definition but should reuse the *common management interface protocol* (CMIP) for management interfaces. At the ATM Forum, the fundamental concept of access networks itself was questioned, and it was suggested that it should also be possible to terminate user signaling within access networks and to provide switching and service-level functions there, in addition to transmission and multiplexing functions.

With so much interest, it is clear that there will be a standard specification for the VB5 interface or interfaces. It is not so clear that there will be universal agreement about its nature.

## Selected Bibliography

European Telecommunications Standards Institute, *Identification of the Applicability of Existing Specifications for a VB5 Interface in an Access Arrangement With Access Networks*, DTR/SPS-03040.

European Telecommunications Standards Institute, *Interfaces at the VB5.1 (or VB5.2) Reference Point for the Support of Broadband or Combined Narrowband and Broadband Access Networks*, DE/SPS-03046.1 (or 03047.1).

McDysan, D. E., and D. L. Spohn, *ATM Theory and Application*, New York: McGraw-Hill, 1995.

# 18 Concluding Remarks

"I don't believe in an afterlife, although I am bringing a change of underwear."

—Woody Allen

This book has grown in the telling. Partly this is a reflection of the developments in access network technology, and partly it is a reflection of the increasing significance of broadband communications. Meanwhile, the V5 interface has developed from one of the many twinkles in Karl-Heinz Stolp's eyes to an accepted global standard.

So where will all this end? Some people believe that in the new millennium simple telephony will be replaced by ATM networks and computer communications. Perhaps they are right, but it may be worthwhile remembering what may have been humanity's first two inventions: fire and the wheel. Although homes may be warmed by electric heaters and vehicles may be supported by cushions of air, fire and the wheel are so fundamental that it is hard to believe that they will ever be fully replaced. The same may be true of simple telephony.

# Acromyms and Abbreviations

**μs** microsecond

**1pC** one per customer

**AAL** ATM adaptation layer

**ABR** available bit rate

**ADPCM** adaptive delta pulse code modulation

**ADSL** asymmetric digital subscriber loop

**AMI** alternative mark inversion

**AMPS** advanced mobile phone service

**ANSI** American National Standards Institute

**APON** ATM PON

**ASE** amplified spontaneous emission

**ATM** asynchronous transfer mode

**BCC** bearer channel connection

**BER** bit error rate

**CAI** common air interface

**CAP** carrierless amplitude/phase

**CATV** cable television

**CBR** constant bit rate

**CD** compact disc

**CDM** code-division multiplexing

**CDMA** code division multiple access

**CMIP** common management interface protocol

**CPE** customer premises equipment

**C/R** command/response

**CRC** cyclic redundancy checksum

**CSA** carrier serving area

**CT1** first-generation cordless telephony

**CT2** second-generation cordless telephony

**CTP** connection termination point

**DDD** directional division duplexing

**DDI** direct dialing in

**DDM** directional division multiplexing

**DECT** digital European cordless telecommunications

**DFE** decision feedback equalization

**DMT** discrete multitone

**DS** digital section

**DSC1800** Digital Communications System at 1800 MHz

**DSL** digital subscriber loop

**DSP** digital signal processing

**DTFM** dual tone multifrequency

**EA** extension address

**EC** European Community

**EDFA** erbium-doped fiber amplifier

**EPR** Einstein-Podolsky-Rosen

**ETSI** European Telecommunications Standards Institute

**FCS** frame check sequence

**FDD** frequency-division duplex

**FDDI** fiber-optic data distribution interface

**FDM** frequency-division multiplexing

**FDMA** frequency division multiple access

**FEXT** far-end crosstalk

**FIR** finite impulse response

**FM** frequency modulation

**FPM** four-photon mixing

**FR** frame relay

**FTS** feeder transmission system

**FTTA** fiber to the apartment

**FTTB** fiber to the business

**FTTC** fiber to the curb

**FTTCab** fiber to the cabinet

**FTTH** fiber to the home

**FWM** four-wave mixing

**GHz** gigahertz

**GSM** Groupe Speciale Mobile

**HDSL** high-speed digital subscriber loop

**HDTV** high-definition television

**HDWDM** high-density wavelength-division multiplexing

**HFC** hybrid fiber/coax

**IIR** infinite impulse response

**ISDN** integrated services digital network

**ISI** intersymbol interference

**ITU** International Telecommunications Union

**Kbps** kilobits per second

**kft** kilofeet

**kHz** kilohertz

**LAN** local-area network

**LED** light-emitting diode

**LLDN** local line distribution network

**LMDS** local multipoint distribution service

**LMWS** licensed millimeter-wave service

**MAN** metropolitan-area network

**Mbps** megabits per second

**MD** mediation device

**MHz** megahertz

**MPEG** Motion Pictures Experts Group

**ms** millisecond

**MUX** multiplexer

**NE** network element

**NEXT** near-end crosstalk

**ODN** optical distribution network

**OFDD** optical frequency-division duplexing

**OFDM** optical frequency-division multiplexing

**OFSK** optical frequency-shift keying

**OLT** optical line termination

**ONU** optical network unit

**OTN** optical transition node

**OS** operations system

**OSI** Open Systems Interconnection

**PABX** private automatic branch exchange

**PC** personal computer

**PCN** personal communications networks

**PIN** p-type intrinsic n-type

**PM** phase modulation

**PON** passive optical network

**POTS** plain old telephone service

**PSD** power spectral density

**PSTN** public service telephony network

**PT** payload type

**QA** Q-adapter

**QAM** quadrature amplitude modulation

**rDS** remote digital section

**RLT** radio line termination

**RNU** radio network unit

**SBS** stimulated Brillouin scattering

**SCM** subcarrier multiplexing

**SDD** space-division duplexing

**SDH** synchronous digital hierarchy

**SDM** space-division multiplexing

**SMDS** switched multimegabit data service

**SNMP** simple network management protocol

**SPM** self-phase modulation

**SRS** stimulated Raman scattering

**TCM** time-compression multiplexing

**TDD** time-division duplex

**TDM** time-division multiplexing

**TDMA** time division multiple access

**TMN** telecommunications management network

**TTP** trail termination point

**VBR** variable bit rate

**VC** virtual channel

**VCC** virtual channel connection

**VCI** virtual channel identifier

**VCL** virtual channel link

**VCR** video cassette recorder

**VDSL** very-high-speed digital subscriber loop

**VoD** video-on-demand

**VP** virtual path

**VPC** virtual path connection

**VPI** virtual path identifier

**VPL** virtual path link

**WAN** wide-area network

**WDD** wavelength-division duplexing

**WDM** wavelength-division multiplexing

**WWW** World-Wide Web

**XPM** cross-phase modulation

# About the Author

Alex Gillespie has a masters degree from Cambridge and a doctorate from Durham, in addition to his first class honors degree from St. Andrews in Scotland. He has worked on both sides of the Atlantic on a number of aspects of telecommunications. Since 1988 he has been at BT Laboratories at Martlesham Heath, and he has been editor of several telecommunications standards. He is chairman of the ETSI SPS3 Working Party on Management, a member of the ECTM group, established at the request of the European Commission to coordinate telecommunication management standards in Europe, and a rapporteur for ITU-T Study Group 4. His e-mail address is gillesat @btlip23.bt.co.uk.

# Index

## *The Artech House Telecommunications Library*

*Vinton G. Cerf, Series Editor*

*Access Networks: Technology and V5 Interfacing*, Alex Gillespie

*Advanced High-Frequency Radio Communications,* Eric E. Johnson, Robert I. Desourdis, Jr., et al.

*Advanced Technology for Road Transport: IVHS and ATT*, Ian Catling, editor

*Advances in Computer Systems Security, Vol. 3*, Rein Turn, editor

*Advances in Telecommunications Networks,* William S. Lee and Derrick C. Brown

*Advances in Transport Network Technologies: Photonics Networks, ATM, and SDH,* Ken-ichi Sato

*An Introduction to International Telecommunications Law,* Charles H. Kennedy and M. Veronica Pastor

*Asynchronous Transfer Mode Networks: Performance Issues, Second Edition,* Raif O. Onvural

*ATM Switching Systems*, Thomas M. Chen and Stephen S. Liu

*Broadband: Business Services, Technologies, and Strategic Impact*, David Wright

*Broadband Network Analysis and Design*, Daniel Minoli

*Broadband Telecommunications Technology*, Byeong Lee, Minho Kang, and Jonghee Lee

*Cellular Mobile Systems Engineering,* Saleh Faruque

*Cellular Radio: Analog and Digital Systems,* Asha Mehrotra

*Cellular Radio: Performance Engineering,* Asha Mehrotra

*Cellular Radio Systems,* D. M. Balston and R. C. V. Macario, editors

*CDMA for Wireless Personal Communications,* Ramjee Prasad

*Client/Server Computing: Architecture, Applications, and Distributed Systems Management*, Bruce Elbert and Bobby Martyna

*Communication and Computing for Distributed Multimedia Systems,* Guojun Lu

*Community Networks: Lessons from Blacksburg, Virginia,* Andrew Cohill and Andrea Kavanaugh, editors

*Computer Networks: Architecture, Protocols, and Software,* John Y. Hsu

*Computer Mediated Communications: Multimedia Applications,* Rob Walters

*Computer Telephone Integration,* Rob Walters

*Convolutional Coding: Fundamentals and Applications,* Charles Lee

*Corporate Networks: The Strategic Use of Telecommunications*, Thomas Valovic

*Digital Beamforming in Wireless Communications,* John Litva, Titus Kwok-Yeung Lo

*Digital Cellular Radio*, George Calhoun

*Digital Hardware Testing: Transistor-Level Fault Modeling and Testing,* Rochit Rajsuman, editor

*Digital Switching Control Architectures*, Giuseppe Fantauzzi

*Digital Video Communications*, Martyn J. Riley and Iain E. G. Richardson

*Distributed Multimedia Through Broadband Communications Services,* Daniel Minoli and Robert Keinath

*Distance Learning Technology and Applications,* Daniel Minoli

*EDI Security, Control, and Audit*, Albert J. Marcella and Sally Chen

*Electronic Mail*, Jacob Palme

*Enterprise Networking: Fractional T1 to SONET, Frame Relay to BISDN*, Daniel Minoli

*Expert Systems Applications in Integrated Network Management,* E. C. Ericson, L. T. Ericson, and D. Minoli, editors

*FAX: Digital Facsimile Technology and Applications, Second Edition,* Dennis Bodson, Kenneth McConnell, and Richard Schaphorst

*FDDI and FDDI-II: Architecture, Protocols, and Performance*, Bernhard Albert and Anura P. Jayasumana

*Fiber Network Service Survivability*, Tsong-Ho Wu

*A Guide to the TCP/IP Protocol Suite*, Floyd Wilder

*Implementing EDI*, Mike Hendry

*Implementing X.400 and X.500: The PP and QUIPU Systems*, Steve Kille

*Inbound Call Centers: Design, Implementation, and Management*, Robert A. Gable

*Information Superhighways Revisited: The Economics of Multimedia*, Bruce Egan

*Integrated Broadband Networks*, Amit Bhargava

*International Telecommunications Management*, Bruce R. Elbert

*International Telecommunication Standards Organizations*, Andrew Macpherson

*Internetworking LANs: Operation, Design, and Management*, Robert Davidson and Nathan Muller

*Introduction to Document Image Processing Techniques*, Ronald G. Matteson

*Introduction to Error-Correcting Codes*, Michael Purser

*An Introduction to GSM,* Siegmund Redl, Matthias K. Weber,Malcom W. Oliphant

*Introduction to Radio Propagation for Fixed and Mobile Communications,* John Doble

*Introduction to Satellite Communication,* Bruce R. Elbert

*Introduction to T1/T3 Networking*, Regis J. (Bud) Bates

*Introduction to Telephones and Telephone Systems, Second Edition,* A. Michael Noll

*Introduction to X.400*, Cemil Betanov

*LAN, ATM, and LAN Emulation Technologies,* Daniel Minoli and Anthony Alles

*Land-Mobile Radio System Engineering*, Garry C. Hess

*LAN/WAN Optimization Techniques*, Harrell Van Norman

*LANs to WANs: Network Management in the 1990s*, Nathan J. Muller and Robert P. Davidson

*Minimum Risk Strategy for Acquiring Communications Equipment and Services*, Nathan J. Muller

*Mobile Antenna Systems Handbook*, Kyohei Fujimoto and J.R. James, editors

*Mobile Communications in the U.S. and Europe: Regulation, Technology, and Markets*, Michael Paetsch

*Mobile Data Communications Systems*, Peter Wong and David Britland

*Mobile Information Systems*, John Walker

*Networking Strategies for Information Technology*, Bruce Elbert

*Packet Switching Evolution from Narrowband to Broadband ISDN*, M. Smouts

*Packet Video: Modeling and Signal Processing*, Naohisa Ohta

*Personal Communication Networks: Practical Implementation*, Alan Hadden

*Personal Communication Systems and Technologies*, John Gardiner and Barry West, editors

*Practical Computer Network Security*, Mike Hendry

*Principles of Secure Communication Systems*, Second Edition, Don J. Torrieri

*Principles of Signaling for Cell Relay and Frame Relay*, Daniel Minoli and George Dobrowski

*Principles of Signals and Systems: Deterministic Signals*, B. Picinbono

*Private Telecommunication Networks*, Bruce Elbert

*Radio-Relay Systems*, Anton A. Huurdeman

*RF and Microwave Circuit Design for Wireless Communications*, Lawrence E. Larson

*The Satellite Communication Applications Handbook*, Bruce R. Elbert

*Secure Data Networking*, Michael Purser

*Service Management in Computing and Telecommunications*, Richard Hallows

*Smart Cards*, José Manuel Otón and José Luis Zoreda

*Smart Highways, Smart Cars*, Richard Whelan

*Super-High-Definition Images: Beyond HDTV*, Naohisa Ohta, Sadayasu Ono, and Tomonori Aoyama

*Television Technology: Fundamentals and Future Prospects*, A. Michael Noll

*Telecommunications Technology Handbook*, Daniel Minoli

*Telecommuting*, Osman Eldib and Daniel Minoli

*Telemetry Systems Design*, Frank Carden

*Teletraffic Technologies in ATM Networks*, Hiroshi Saito

*Toll-Free Services: A Complete Guide to Design, Implementation, and Management*, Robert A. Gable

*Transmission Networking: SONET and the SDH*, Mike Sexton and Andy Reid

*Troposcatter Radio Links*, G. Roda

*Understanding Emerging Network Services, Pricing, and Regulation*, Leo A. Wrobel and Eddie M. Pope

*Understanding GPS: Principles and Applications*, Elliot D. Kaplan, editor

*Understanding Networking Technology: Concepts, Terms and Trends*, *Mark Norris*

*UNIX Internetworking, Second edition*, Uday O. Pabrai

*Videoconferencing and Videotelephony: Technology and Standards*, Richard Schaphorst

*Wireless Access and the Local Telephone Network*, George Calhoun

*Wireless Communications in Developing Countries: Cellular and Satellite Systems*, Rachael E. Schwartz

*Wireless Communications for Intelligent Transportation Systems*, Scott D. Elliot and Daniel J. Dailey

*Wireless Data Networking*, Nathan J. Muller

*Wireless LAN Systems*, A. Santamaría and F. J. López-Hernández

*Wireless: The Revolution in Personal Telecommunications*, Ira Brodsky

*Writing Disaster Recovery Plans for Telecommunications Networks and LANs*, Leo A. Wrobel

*X Window System User's Guide*, Uday O. Pabrai

*For further information on these and other Artech House titles, contact:*

Artech House
685 Canton Street
Norwood, MA 02062
617-769-9750
Fax: 617-769-6334
Telex: 951-659
email: artech@artech-house.com

Artech House
Portland House, Stag Place
London SW1E 5XA England
+44 (0) 171-973-8077
Fax: +44 (0) 171-630-0166
Telex: 951-659
email: artech-uk@artech-house.com

WWW: http://www.artech-house.com